Recycling and Regeneration of Spent Lithium-Ion Batteries

Recycling and Regeneration of Spent Lithium-Ion Batteries

Technologies, Applications, and Sustainable Futures

Edited by

Peng Ge

Yue Yang

WILEY-VCH

Editors

Prof. Peng Ge and Prof. Yue Yang
School of Minerals Processing and Bioengineering,
Central South University, Changsha,
410083,
China

Cover Design: Wiley
Cover Image: CHRISTOPH BURGSTEDT/
SCIENCE PHOTO LIBRARY/Getty Images

Typesetting: Lumina Datamatics, Madurai, India

Library of Congress Card No.: applied for

British Library Cataloguing-in-Publication Data
A catalogue record for this book is available from the British Library.

Bibliographic information published by the Deutsche Nationalbibliothek The Deutsche Nationalbibliothek lists this publication in the Deutsche Nationalbibliografie; detailed bibliographic data are available on the Internet at <http://dnb.d-nb.de>.

Print ISBN 9783527356669
ePDF ISBN 9783527857104
ePub ISBN 9783527857098
oBook ISBN 9783527857111

C9783527356669_230626

Printed and bound by CPI Group (UK) Ltd,
Croydon, CR0 4YY

Contents

List of Contributors

Xiaoyu Cao
School of Chemistry and Chemical Engineering
Henan University of Technology
Zhengzhou
China

Runxuan Chen
School of Minerals Processing and Bioengineering
Central South University
Changsha
China

Xizhuo Chen
School of Minerals Processing and Bioengineering
Central South University
Changsha
China

Zeyu Dong
School of Minerals Processing and Bioengineering
Central South University
Changsha
China

Peixiang Gao
School of Environmental Engineering
Henan University of Technology
Zhengzhou
China

Peng Ge
School of Minerals Processing and Bioengineering
Central South University
Changsha
China

Shuaiqi Gong
Shanghai Key Laboratory of Materials Protection and Advanced Materials in Electric Power
Shanghai University of Electric Power
Shanghai
China

Binglei Jiao
Soochow Institute of Energy and Material Innovations
College of Energy
Soochow University
Suzhou
China

Tiyu Jiao
Soochow Institute of Energy and Material Innovations
College of Energy
Soochow University
Suzhou
China

Mengqi Kong
Soochow Institute of Energy and Material Innovations
College of Energy
Soochow University
Suzhou
China

Hai Lei
School of Minerals Processing and Bioengineering
Central South University
Changsha
China

Jiexiang Li
School of Minerals Processing and Bioengineering
Central South University
Changsha
China

Xiangjin Lu
School of Minerals Processing and Bioengineering
Central South University
Changsha
China

Zihao Lu
School of Minerals Processing and Bioengineering
Central South University
Changsha
China

Congrui Ouyang
College of Materials Science and Engineering
Chongqing University
Chongqing
China

Xuejing Qiu
School of Environmental Engineering
Henan University of Technology
Zhengzhou
China

Run Ren
School of Minerals Processing and Bioengineering
Central South University
Changsha
China

Penghui Shi
Shanghai Key Laboratory of Materials Protection and Advanced Materials in Electric Power
Shanghai University of Electric Power
Shanghai
China

Gaoyun Tan
School of Minerals Processing and Bioengineering
Central South University
Changsha
China

Jin Wang
College of Materials Science and Engineering
Chongqing University
Chongqing
China

Yunpeng Wen
School of Minerals Processing and Bioengineering
Central South University
Changsha
China

Lingling Xie
School of Environmental Engineering
Henan University of Technology
Zhengzhou
China

Panpan Xu
Soochow Institute of Energy and
Material Innovations
College of Energy
Soochow University
Suzhou
China

Yue Yang
School of Minerals Processing and
Bioengineering
Central South University
Changsha
China

Zhichao Yang
Soochow Institute of Energy and
Material Innovations
College of Energy
Soochow University
Suzhou
China

Wenhao Yu
College of Materials Science and
Engineering
Chongqing University
Chongqing
China

Zihao Zeng
School of Minerals Processing and
Bioengineering
Central South University
Changsha
China

Jiale Zhang
School of Minerals Processing and
Bioengineering
Central South University
Changsha
China

Quan Zhang
School of Minerals Processing and
Bioengineering
Central South University
Changsha
China

Yimeng Zhang
School of Environmental Engineering
Henan University of Technology
Zhengzhou
China

Hanyu Zhou
School of Minerals Processing and
Bioengineering
Central South University
Changsha
China

Chao Zhu
School of Minerals Processing and
Bioengineering
Central South University
Changsha
China

Limin Zhu
School of Chemistry and Chemical
Engineering
Henan University of Technology
Zhengzhou
China

Preface

Energy, as the main resource, served as an important role for the development of vital technologies in society. However, considering the serious consumption of fossil energy with disappointing environmental pollution, new-type clean energy has been devoted to plenty of attention, such as wind energy, solar energy, etc. As a prediction, from 2025 to 2030, the total installed capacity of global clean energy would be increasing with an additional 4,600 GWh. Obviously, clean energy resources have been regarded as one of the main energy sources. But it should not be neglected that they still suffer from intermittent and unsustainable abilities. Lithium-ion batteries (LIBs), as the vital energy conversion systems, have obtained numerous achievements. Nowadays, they have been widely applied in electro-mobiles, energy-storage stations, and so on. Certainly, based on the utilization of new-type energy-storage technologies, a green and clean environment was gradually formed. Meanwhile, the 2019 Nobel Prize in Chemistry was awarded to John B. Goodenough, M. Stanley Whittingham and Akira Yoshino, for their contributions to the development of LIBs. Their work has created a rechargeable world for humanity, bringing about numerous progresses in other emerging fields, such as unmanned aircraft systems and intelligent robots. According to the forecast, by 2030, the combined shipment volume of global battery power and energy storage batteries is expected to reach 5,154 GWh, which is equivalent to 25.77 million tons of LIBs. However, some discussions have been carried out about the clean properties of LIBs at all process from raw materials to spent batteries through the tracing of vital elements and pollutions. As-known, constrained by the limited usage loop (8–10 years), a series of spent batteries would be resulted. Without suitable recycling treatments, serious elements waste and pollution would be noted. Thus, exploring the highly effective recycling strategies of spent LIBs is important for the establishment of green and clean energy-storage systems.

Based on the previous reports, spent LIBs are composed of rich metal elements, such as lithium, cobalt, nickel, and so on. Thus, they have been deemed as the valuable "urban mines." Differing from raw minerals, the structure of LIBs was designed by complex manual processes, such as cylindrical batteries, blade batteries, and so on. Moreover, owing to the residual energy and complex organic/inorganic components, they bring about the high complexity and inferior safety of recycling operations. Thus, all-process chain of recycling manners has been explored, containing pre-treatments, cascade utilizations, pyrometallurgy-hydrometallurgy elements-extracting manners, direct regeneration, pollution inhibition, and all-life process evaluations. Herein, the current status of relative technologies is illustrated, accompanying with clear mechanism exploration. More importantly, their development tendencies are also displayed through the comparison of relative technologies. Herein, the book consists of eight chapters. For Chapter 1, the brief introduction of recycling LIBs is displayed through the illustration of production volume and distribution, etc. For Chapter 2, the pretreatment process contains battery structure and classification, detection and evaluation, safe discharge, disassembly and crushing, and material sorting; for Chapter 3, cascade utilization contains sorting of retired LIBs, sorting and reconfiguration, etc. For Chapter 4, the relative element-extracting manners are displayed through the distinguishing of pyrometallurgy and hydrometallurgy; for Chapter 5, based on diverse main electrode materials, their direct regenerations are clearly separated and discussed. For Chapter 6, the relative pollution inhibition and all-life evaluation are illustrated. For Chapter 7, based on different technologies, the different development trends and mainstream directions of technologies are shown. For Chapter 8, the final conclusion is shown.

This book is prepared with the support of many peer researchers who made important contributions to the final version of this book. Chapter 2 is authored by Professors Peng Ge and Yue Yang, Chapter 3 is authored by Dr. Shuaiqi Gong and Professor Penghui Shi, and Chapter 4 is authored by Dr. Xuejing Qiu and Professors Xiaoyu Cao, Limin Zhu, and Lingling Xie. Chapter 5 is authored by Associate Professor Panpan Xu, and Chapter 6 is authored by Associate Professor Wenhao Yu. The other chapters are mainly authored by Professors Peng Ge and Yue Yang, and they also completed the compilation and revision of this book.

Finally, the authors would like to acknowledge the financial support of the National Natural Science Foundation of China (52534010).

1

Introduction

Hai Lei, Chao Zhu, Jiexiang Li, Hanyu Zhou, Xiangjin Lu, Xizhuo Chen, Yue Yang

1.1 Production Volume and Distribution of Spent LIBs

Spent lithium-ion batteries (LIBs) exhibit dual characteristics of type differentiation and regional concentration in production volume and distribution.[1–4] In Figure 1.1, as the core source, the end-of-life (EoL) volume of power batteries in China is projected to reach 820,000 tons in 2025. Among them, NCM batteries (accounting for 55%) display a relatively short cycle life (5–8 years) and are concentrated in passenger vehicle-dense areas such as the Yangtze River Delta and the Pearl River Delta. In contrast, LFP batteries (accounting for 45%) are experiencing rapid growth in commercial vehicle-dense areas such as the Chengdu-Chongqing and Beijing-Tianjin-Hebei regions, owing to their extended lifespan of 8–10 years. And, it is anticipated that by 2030, the proportion of LFP batteries in the total EoL power battery stream will rise to 60%. Moreover, the retired consumer electronic batteries contribute about 150,000 tons annually, dominated by LCO batteries (70%), which has a high degree of dispersion, resulting in a formal recycling rate of less than 20%. These batteries are mainly distributed in the Pearl River Delta electronics manufacturing hub (accounting for 35% of the national total) and the cascade-utilization project areas in Jiangxi. As the main force for cascade utilization, the market scale of energy-storage batteries is forecasted to reach RMB 60 billion, with over 85% LFP batteries. Moreover, 40% cascade-utilization batteries are collectively absorbed by northwestern photovoltaic energy storage zones (Qinghai/Ningxia) and north China grid-scale energy-storage areas across the nation. For specialized batteries (such as NCM and lithium metal batteries), their annual EoL volume is around 30,000 tons, concentrated in port cities like Tianjin and Qingdao, as well as high-end recycling project areas in Hunan.

Furthermore, the geographical distribution pattern for recycling batteries is as follows: first, the power batteries recycling network is concentrated in major automotive industry clusters (e.g., Chongqing plans to achieve 90% coverage); second,

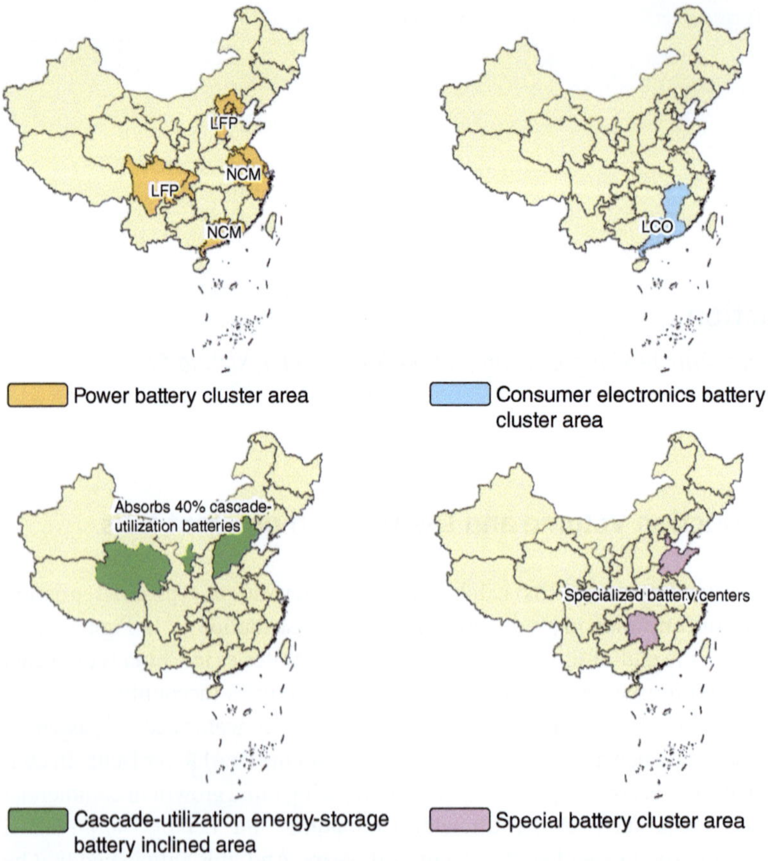

Figure 1.1 Geographical distribution of spent LIBs and recycling networks in China.

consumer electronic batteries recycling primarily relies on the Pearl River Delta electronics manufacturing cluster; third, cascade-utilization energy-storage batteries recycling is preferentially directed toward the northwest energy bases.[5] In addition, significant differentiation exists in technological recycling pathways: owing to their high-value cobalt and nickel content, NCM batteries are prioritized for recycling and regeneration (hydrometallurgical recovery rates $\geq$ 98%).[6, 7] Conversely, LFP batteries are primarily destined for cascade utilization.[8, 9] Meanwhile, policy drivers have emerged as critical variables. Initiatives such as the Carbon Border Adjustment Mechanism (CBAM) in the European Union (EU) and regional pilot programs (e.g., the battery traceability system in Chongqing) in China are accelerating industry standardization. Therefore, given the projected continuous increase in EoL battery volumes, the synergistic optimization of geographical deployment strategies and technological pathways will be the cornerstone for achieving highly efficient resource circulation.

1.2 Strategic Significance of Spent LIBs Recycling

As shown in Figure 1.2, the recycling of spent LIBs holds multidimensional strategic importance for global sustainable development, encompassing resource security, environmental governance, climate action, and industrial restructuring.[1, 10] From a resource security perspective, critical metals such as lithium (Li), cobalt (Co), and nickel (Ni) are geographically concentrated in a few countries (e.g., Chile for Li, the Democratic Republic of Congo [DRC] for Co), posing significant geopolitical risks. Therefore, the recycling industry enhances supply chain resilience through material regeneration. According to the data from the International Energy Agency (IEA), recycled materials are projected to meet 35% of global Li demand and 50% of Co demand by 2040, meanwhile reducing the carbon footprint of mining to 20% of virgin ores and cutting water consumption by 90%.[11, 12]

In terms of environmental governance, recycling systems mitigate the cross-border migration risks of hydrofluoric acid (HF) generated by $LiPF_6$ decomposition. Moreover, according to UN statistics, it can also prevent approximately 500,000 tons of electronic

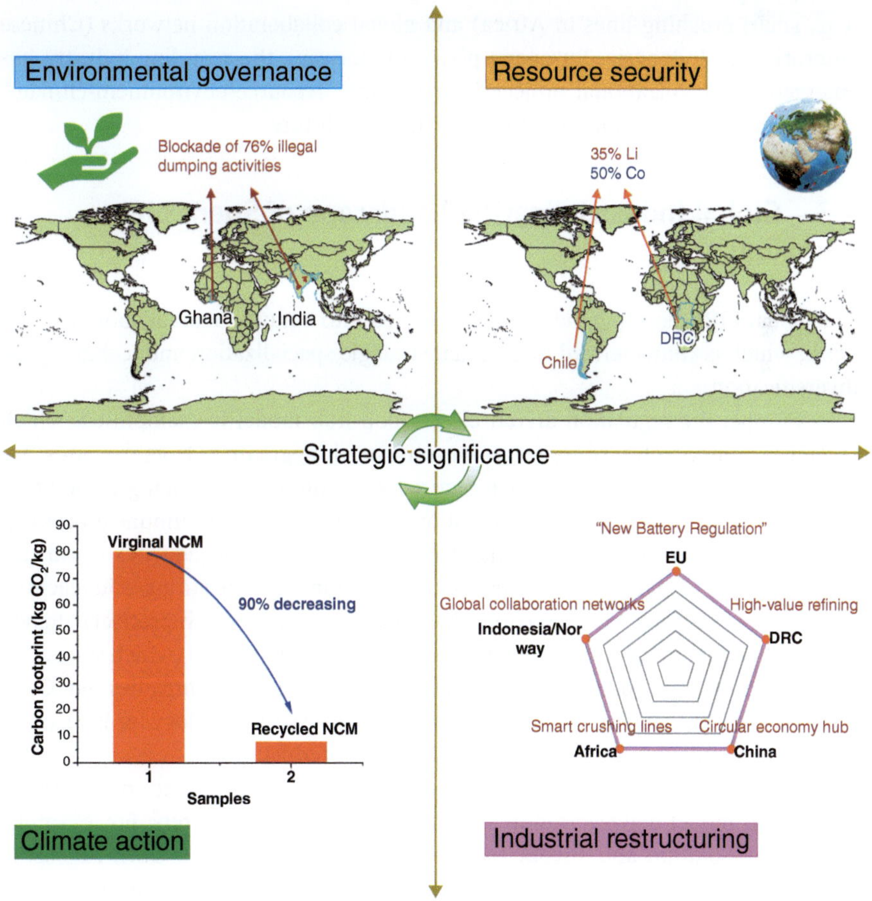

Figure 1.2 Global impact of LIBs recycling.

batteries (accounting for 76% of the total) from entering informal dismantling networks (such as Ghana, India, and other regions) annually, thereby avoiding heavy metal contamination of 75 million cubic meters of land.

For climate action, recycling serves as a pivotal lever for carbon neutrality. The carbon footprint of recycled NCM materials is about 8 kg CO_2/kg, which is 90% lower than that of virgin materials (80 kg CO_2/kg). By 2040, 530 million tons of annual CO_2 emissions (equivalent to the UK's annual emissions) would be reduced by LIBs recycling in the global transportation sector. Additionally, cascade utilization of batteries can lower the cost of photovoltaic energy storage systems by 40%, accelerating renewable energy adoption.

Regarding industrial restructuring, recycling is reshaping global value chains. Specifically, the "New Battery Regulation" in the EU mandates a minimum 12% recycled Co and Ni content by 2030, driving cost reductions in cell manufacturing for developed countries.[13] And, resource-rich nations like the DRC are transitioning to high-value refining industries, while China is upgrading from the "world's factory" to a "circular economy hub" with a projected CNY 200 billion recycling industry by 2030. In conclusion, through modular technology dissemination (e.g., smart crushing lines in Africa) and global collaboration networks (Chinese enterprises in Indonesia, European plants in Norway), the recycling industry has emerged as a civilizational project for achieving "resource-environment-climate" synergy and will underpin humanity's sustainable future.

1.3 Global Industrialization Landscape of Spent LIBs Recycling

The global LIBs recycling industry has evolved into a multipolar collaborative network, which is characterized by regional strategic specialization and technological diversification.

In Europe, the regulation-driven closed-loop hub model is exemplified, which leverages policy-technical synergy for industrial upgrading. Notable cases are as follows: Tozero in Germany has secured €11 million in financing to build an industrial-scale recycling plant (operational in 2026) with an annual processing capacity of 30,000 tons. And this plant features an innovative aqueous-based carbonation process that eliminates secondary pollution from the strong-acid leaching process. Besides, for CATL, the Hungary-based recycling project is precisely aligned with the requirements of the "New Battery Regulation" in the EU (which states that the Li recovery rate should be no less than 80%). Moreover, companies like Cylib have set industry benchmarks with oxygen-free pyrolysis technology for EU carbon footprint compliance, accompanied by 40% carbon reduction.

In North America, the development of the recycling industry relies on the vertical integration model between automakers and battery manufacturers. For example, LG Energy Solution and Toyota Tsusho have jointly established GMBI, focusing on the "battery-to-battery" closed-loop production, which is expected to process 13,500 tons of spent LIBs by 2026. Meanwhile, they leverage the tax incentives in the Inflation Reduction Act to avoid the EU's carbon barriers.

Asia-Pacific, led by China, drives large-scale and networked expansion of technology. EVE Energy, in collaboration with GEM and Huayou Cobalt, has built a global LIBs recycling network platform, planning to reach an annual production capacity of 120,000 tons by 2030, covering 246 sites across 30 countries, accompanied by the integration of "physical-data-value" flow through blockchain technology. In contrast, VinFast in Vietnam has cooperated with BatX Energies in India to optimize the local supply chain in Southeast Asia.

For the Middle East, the rise of emerging hubs is accelerating. The United Arab Emirates has invested $10.8 million to build the region's first LIBs recycling plant in Abu Dhabi, mainly dealing with batteries from photovoltaic energy storage and electric vehicles, aiming to become the core node of the circular economy in the Middle East.

In conclusion, the collaborative cooperation among these four regions is driving the global resource closed-loop to evolve from "single-point breakthrough" to "system-level reconstruction."

1.4 Full-chain Technological Process for High-value Recycling of Spent LIBs

The high-value recycling of spent LIBs adheres to the strategy that cascade utilization comes first and material regeneration subsequently, comprising the following critical stages in Figure 1.3. At the beginning, the pack-level battery state is detected at the preprocessing and cascade utilization screening stage. Specifically, battery packs with parameters such as capacity, internal resistance, and power that meet the requirements for direct cascade utilization are selected for direct application. In contrast, unqualified battery packs are processed by intelligent disassembly technology to obtain battery cells, and then rapid detection and state evaluation are carried out at the cell level.

In the cascade utilization and safe discharge stage, the qualified cells are classified and recombined into battery packs with high consistency according to the

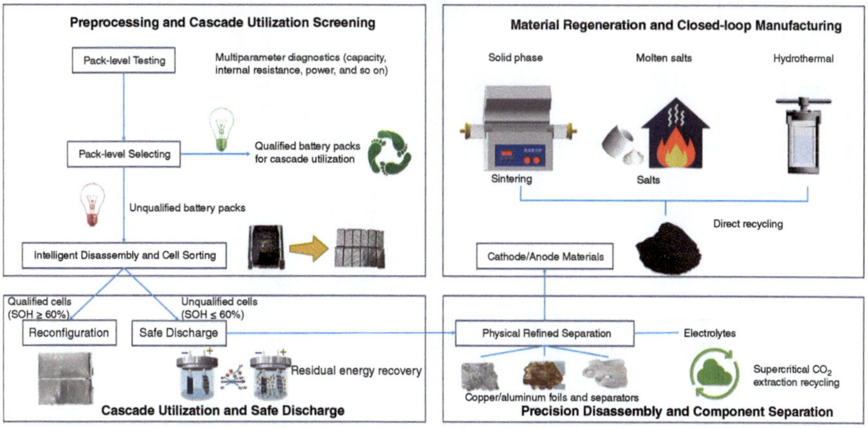

Figure 1.3 Full-chain recycling process flowchart.

requirements of cascade utilization scenarios for secondary use. For the cells that cannot be cascaded, the remaining energy is completely recovered through intelligent discharge technology, while ensuring the safety of subsequent operations.

After complete discharge, the internal components of the battery are completely classified and recycled through precision disassembly technology. Components that can be directly recycled, such as copper/aluminum foils and separators. For products that have caused secondary pollution (such as electrolytes), clean processes should be adopted for recycling or harmless treatment.

In the last stage, for the recycling of cathode/anode materials, according to the type of battery materials and the judgment of the material failure mechanism during the detection process, the material performance is directly regenerated through specific recycling technologies such as solid phase, molten salts, and hydrothermal methods. The regenerated battery materials are reused for the assembly of new LIBs, and the assembled battery packs are reused in various energy storage fields. Thus, high-value circular utilization of LIBs throughout their life cycle is achieved.

1.5 Overview of Global Progress in LIBs Recycling

With the growing global emphasis on clean energy and sustainable development, as efficient and environmentally friendly energy storage devices, LIBs have been widely adopted in electric vehicles, consumer electronics, and energy storage systems.[14] According to a report by the IEA in Figure 1.4, the global EV stock reached 27 million units in 2022, with China and Europe collectively contributing the majority (85%) of global EV sales. Under the Net Zero Emissions by 2050 Scenario, the global EV stock is projected to exceed 350 million units by 2030. Concurrently, the number of EoL EVs is anticipated to grow significantly in the coming years.

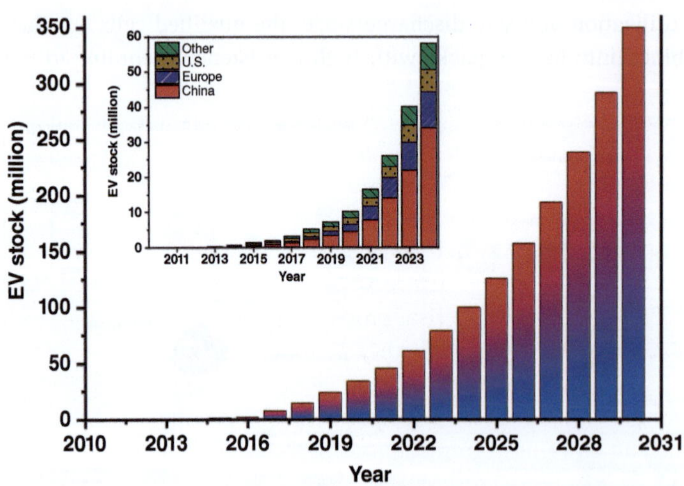

Figure 1.4 EV stock in major economies from 2010 to 2022 and projected stock for the next decade under the Net Zero Emissions by 2050 Scenario.

However, with the large-scale utilization of LIBs, the problem of their postretirement disposal has become increasingly prominent. Improper handling of spent LIBs not only leads to resource wastage but also poses significant environmental pollution risks.[15] Globally, the development of spent LIBs recycling technologies exhibits diversity and rapid advancement. Countries are adopting varying strategies concerning policy support, technological research, and industrial chain development, driving continuous innovation and the industrialization of recycling processes.

The EU holds a leading position in the recycling of EoL LIBs, with its success largely attributable to stringent regulations and policy support.[16] Through directives such as the "Battery Directive" and the "Circular Economy Action Plan," the EU explicitly defines the recycling responsibilities of battery producers and importers, establishing specific recycling targets and timelines. Furthermore, the EU promotes research, development, and demonstration applications of recycling technologies through financial support and scientific projects. For instance, the "pyrometallurgical-hydrometallurgical" combined process developed by Umicore in Belgium has achieved a comprehensive Li recovery rate of over 90%, with an annual processing capacity of 70,000 tons. The EU-funded "Battery 2030+" project aims to develop next-generation battery recycling technologies to enhance recovery efficiency and resource utilization. Complementing these efforts, the newly enacted EU Battery Regulation (2023) mandates minimum recovery rates of 70% for Li and 95% for Co/Ni by 2030.[17]

Japan has also made significant progress in the development of technologies and equipment for the recycling of EoL LIBs. Through regulations such as the "Law for the Promotion of Effective Utilization of Resources" and the "Small Electrical and Electronic Equipment Recycling Law," comprehensive battery recycling systems are established by the Japanese government. Japanese companies like Toyota, Panasonic, and Nissan have invested heavily in battery recycling technology research and equipment manufacturing. For example, Toyota has developed high-efficiency battery dismantling and material separation technologies, while Panasonic focuses on the research and development of automated recycling equipment, enhancing the efficiency and safety of the recycling process. Besides, in collaboration with JX Metal, Honda has jointly developed an automated single dismantling line, where processing capacity achieves 200 batteries per hour, with a metal purity of 99.95%. And through the "Battery to Battery" program, the Japanese Ministry of Economy, Trade and Industry aims to achieve a 50% recycling rate of vehicle battery materials by 2025.[18]

For the United States, strong research capabilities and industrialization capacities are demonstrated in the field of LIBs recycling.[19, 20] The US government has regulated the battery recycling and disposal process through laws and regulations such as the Resource Conservation and Recovery Act. And supported by the funding for research projects, the US Department of Energy has promoted the innovation of recycling technologies. For instance, Argonne National Laboratory in the United States has developed an efficient hydrometallurgical recycling technology that can extract valuable metals from retired batteries. Additionally, companies like Tesla are actively expanding their battery recycling business and have developed automated recycling production lines, enhancing recycling efficiency and resource utilization.

The closed-loop recycling system at Tesla's Gigafactory Nevada achieves a 92% metal recovery rate and displays a processing cost per ton that is 30% lower than the industry average. The cathode material repair technology developed by the ReCell Center in the United States can retain 90% of the original material structure, and the energy consumption is 50% lower than that of traditional processes.[21]

As the largest producer and consumer of LIBs in the world, China has also made remarkable progress in the field of retired battery recycling in recent years.[22–24] The Chinese government has defined the recycling responsibilities of battery manufacturers, sellers, and consumers through policies such as the "Interim Measures for the Recycling and Utilization of Power Batteries in New Energy Vehicles" and established recycling service networks. Moreover, China has implemented substantial policy support through national science and technology programs and industrial funds to promote the research, development, and industrialization of battery recycling technologies. Leading companies such as CATL and BYD have actively expanded their battery recycling businesses, developing advanced physical and chemical recycling technologies that significantly contribute to the improvement of the recycling industry chain. As of 2023, there are 156 battery recycling enterprises certified by the Ministry of Industry and Information Technology, with an annual processing capacity of over two million tons. Notably, Brunp Recycling (a CATL subsidiary) has established the world's largest hydrometallurgical recycling facility, boasting an annual processing capacity of 120,000 tons. Additionally, GEM Co., Ltd. has developed an innovative multistage crushing-sorting-leaching process that achieves remarkable recovery rates of 98.5% for Ni/Co and over 85% for Li_2CO_3.

In general, the evolution of EoL LIBs recycling systems demonstrates three defining characteristics: policy-driven support, technological innovation, and industrial chain optimization. Countries have promoted the continuous progress of recycling technologies and the industrialization process through regulations, policy support, and scientific research projects. In the future, with the further maturation of technologies and the gradual improvement of policies, the recycling of EoL LIBs will play an increasingly important role globally, making greater contributions to sustainable resource utilization and environmental protection.

References

1 M. Chen, X. Ma, B. Chen, R. Arsenault, P. Karlson, N. Simon and Y. Wang, *Joule*, 2019, **3**, 2622–2646.

2 L. L. Driscoll, A. Jarvis, R. Madge, E. H. Driscoll, J.-M. Price, R. Sommerville, F. S. Tontini, M. Bahri, M. Miah, B. L. Mehdi, E. Kendrick, N. D. Browning, P. K. Allan, P. A. Anderson and P. R. Slater, *Joule*, 2024, **8**, 2735–2754.

3 S. Zheng, T. Chen, Y. Fang, C. He, H. Duan, S. Ren and C. C. Xu, *Resources Chemicals and Materials*, 2024, **3**, 188–229.

4 F. Gu, J. Guo, X. Yao, P. A. Summers, S. D. Widijatmoko and P. Hall, *Journal of Cleaner Production*, 2017, **161**, 765–780.

5 Z. Qu, W. Huang, Z. Zhou, X. Lei and D. Han, *Energy Sources, Part A: Recovery, Utilization, and Environmental Effects*, 2022, **44**, 7573–7584.

6 M. Yoon, J.-S. Park, W. Chen, Y. Huang, T. Dai, Y. Lee, J. Shin, S. Lee, Y. Kim, D. Lee, D. Shin, J. Cho, Y. Dong and J. Li, *Energy & Environmental Science*, 2025, **18**, 5902–5912.

7 M. L. Nisbet, D. Luong, E. Allen, S. Park, T. L. Kinnibrugh, J. E. Stubbs, P. J. Eng, J. T. Vaughey and T. T. Fister, *Advanced Energy Materials*, 2025, DOI: 10.1002/aenm.202500698.

8 P. Xu, Q. Dai, H. Gao, H. Liu, M. Zhang, M. Li, Y. Chen, K. An, Y. S. Meng, P. Liu, Y. Li, J. S. Spangenberger, L. Gaines, J. Lu and Z. Chen, *Joule*, 2020, **4**, 2609–2626.

9 X. Chang, M. Fan, B. Yuan, W.-H. He, C.-F. Gu, C. Li, Q. Meng and Y.-G. Guo, *Angewandte Chemie International Edition*, 2024, **63**, e202406557.

10 C. Liu, J. Lin, H. Cao, Y. Zhang and Z. Sun, *Journal of Cleaner Production*, 2019, **228**, 801–813.

11 E. M. Bibra, E. Connelly, S. Dhir, M. Drtil, P. Henriot, I. Hwang, J.-B. Le Marois, S. McBain, L. Paoli and J. Teter, 2022. Securing supplies for an electric future. https://www.iea.org/reports/global-ev-outlook-2022.

12 T.-Y. Kim, T. Gould, S. Bennet, F. Briens, A. Dasgupta, P. Gonzales, A. Gouy, G. Kamiya, M. Karpiniski and J. Lagelee, International Energy Agency, Washington, DC, 2021, 70–71.

13 A. Shqairat, S. Liarte, P. Marange, C. Nuur and A. Chagnes, *Management of Environmental Quality: An International Journal*, 2025, **36**, 155–182.

14 E. Mossali, N. Picone, L. Gentilini, O. Rodrìguez, J. M. Pérez and M. Colledani, *Journal of Environmental Management*, 2020, **264**, 110500.

15 J. Wesselkämper, T. P. Hendrickson, S. Lux and S. von Delft, *Environmental Science & Technology*, 2025, **59**, 15751–15765.

16 C. Giosuè, D. Marchese, M. Cavalletti, R. Isidori, M. Conti, S. Orcioni, M. L. Ruello and P. Stipa, *Sustainability*, 2021, **13**, 11154.

17 H. Zhou, Y. Yang, W. Li, J. McKechnie, S. Thiede and P. Wang, *One Earth*, 2024, **7**, 1288–1300.

18 Wang Chang and Xu Jian, World Nonferrous Metals, 2023, 62–65

19 J. Dunn, A. Kendall and M. Slattery, *Resources, Conservation and Recycling*, 2022, **185**, 106488.

20 L. Gaines, K. Richa and J. Spangenberger, *MRS Energy & Sustainability*, 2018, **5**, E14.

21 J. Zhou, X. Zhou, W. Yu, Z. Shang and S. Xu, *Electrochemical Energy Reviews*, 2024, **7**, 13.

22 X. Lou, P. Yan, B. Jiao, Q. Li, P. Xu, L. Wang, L. Zhang, M. Cao, G. Wang, Z. Chen, Q. Zhang and J. Chen, *Nature Communications*, 2024, **15**, 2730.

23 J. Ma, J. Wang, K. Jia, Z. Liang, G. Ji, H. Ji, Y. Zhu, W. Chen, H.-M. Cheng and G. Zhou, *Nature Communications*, 2024, **15**, 1046.

24 H. Wang, X. Geng, L. Hu, J. Wang, Y. Xu, Y. Zhu, Z. Liu, J. Lu, Y. Lin and X. He, *Nature Communications*, 2024, **15**, 1634.

2

Pretreatment

Shuaiqi Gong, Penghui Shi

2.1 Battery Structure and Classification

2.1.1 The Core Structure of Cell

Before delving into the shape classification of lithium-ion batteries, understanding their core electrochemical structure is fundamental. Regardless of the final packaging shape, a typical lithium-ion cell contains the following key internal components. For cathodes, it is usually composed of lithium metal oxides (such as lithium cobalt oxide [$LiCoO_2$], lithium iron phosphate [$LiFePO_4$], NMC, and NCA) or phosphate coated on aluminum foil current collector. During discharge, lithium ions are released from the cathode material and inserted into the cathode; the charging process is reversed.[1] For anodes, it is usually composed of graphite, silicon-based material, or lithium titanate (LTO) coated on copper foil current collector. Receive and store lithium ions migrated from the anode.[2] For separators, a polymer film (such as polyethylene [PE], polypropylene [PP], or its composite) with a microporous structure (allowing lithium ions to pass through). Its core role is to physically isolate the cathode and anode, prevent internal short circuits, and ensure ion conduction. The performance of the separator (porosity, thickness, thermal stability, and mechanical strength) is critical to battery safety and performance. For electrolytes, they are about a mixture of organic solvents (such as ethylene carbonate [EC], dimethyl carbonate [DMC], etc.) in which lithium salts (such as $LiPF_6$ and LiTFSI) are dissolved.[3] It acts as an ion transport medium, conducting lithium ions between cathodes and anodes. Some batteries use solid or semisolid electrolytes. For current collectors, aluminum and copper foil are always selected as the collecting cathode and anode, respectively. They are responsible for conducting the current generated by the active material to the external terminals of the battery.[4] These core components are assembled into electrode groups (jelly roll or stack) through precision winding or stacking processes, and then encapsulated in specific casings or packaging materials,

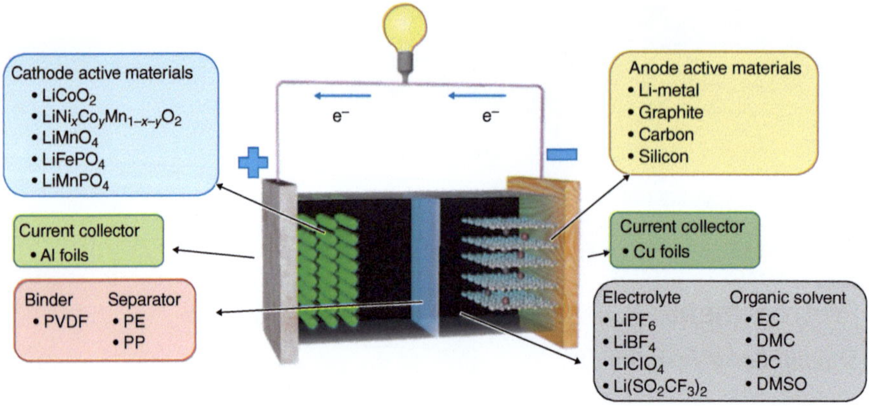

Figure 2.1 Diagram of different components of a lithium-ion battery. *Source*: [6] / with permission of Springer Nature.

injected with electrolyte, and subjected to formation, aging, and other processes, and finally form the lithium-ion single-cell batteries we see in different shapes as shown in Figure 2.1.[5]

2.1.2 Classified by Shape Structure: Cylindrical, Square, and Soft Bag (Polymer)

Lithium-ion battery shape structure is one of its most intuitive features, directly affecting the battery pack space utilization (group efficiency), thermal management design, mechanical strength, cost, and final application scenarios. Mainstream is divided into three categories:[7]

2.1.2.1 Cylindrical Cell

1. **Typical models and evolution**

   In the early days, 18650 (18 mm diameter × 65.0 mm length) was widely used in laptops, power tools, and early electric vehicles (EVs) (such as Tesla Roadster and Model S/X). Subsequently, with the continuous exploration, 21700 (21 mm diameter × 70.0 mm length)—energy density and capacity are higher than 18650, becoming an important choice for current high-end EVs (such as Tesla Model 3/Y) and energy storage. Nowadays, 4680 (46 mm diameter × 80 mm length) displays a significant size increases, mainly resulting from tab-less/full-tab technology (reducing internal resistance, improving heat dissipation, and enhancing rate performance). The internal structure and composition of a cylindrical battery are illustrated in Figure 2.2. The design is beneficial for further increasing energy density and reducing costs (fewer individual cells, simplified assembly). Supported by Tesla, other manufacturers are also following similar specifications. About main application scenarios, EVs (especially Tesla), electric two-wheelers, electric tools, laptops, portable electronic devices, and home energy storage systems could be noted.

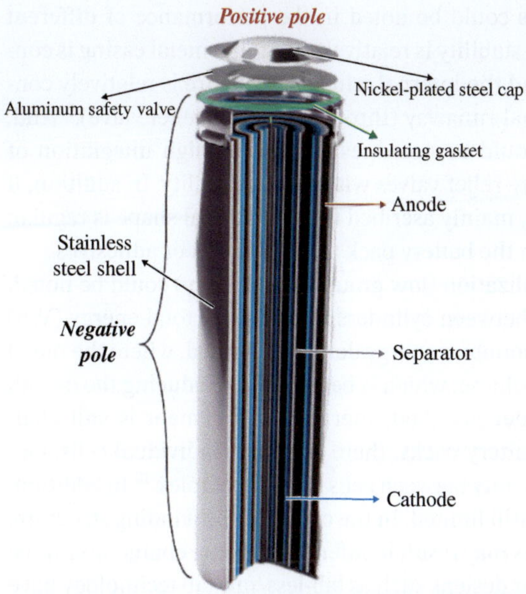

Figure 2.2 The structure and composition of cylindrical batteries. *Source*: [11] / with permission of Elsevier.

2. **About architectural feature**
   A strong metal (usually steel, sometimes high-end or power battery aluminum) cylindrical housing (can) is used. And, the internal electrode group (cathode, separator, and anode) is tightly wound into cylinders in a winding manner (jelly roll). Meanwhile, the case also serves as an electrical connection for the anode current collector (anode ear welded or crimped to the bottom of the case). Herein, the cathode is led out through a metal cap, which usually integrates a safety relief device (such as current interrupt device [CID], positive temperature coefficient [PTC] element, or bursting disc). The device is activated when the internal pressure or temperature of the battery increases abnormally, cutting off the current or releasing the pressure to prevent explosion.[8] Moreover, the cover and housing are insulated and sealed through an insulating gasket (usually using laser welding or mechanical crimping).
3. **About manufacturing process characteristics**
   Highly standardized and automated processes are always applied due to high production maturity and consistency. And, the winding process displayed high efficiency. Meanwhile, the processes of metal casing stamping, winding, liquid injection, sealing, formation, and aging are highly integrated.
4. **Advantages and disadvantage**
   In order to improve high mechanical strength, the metal casing is always used, offering excellent impact and crush resistance. Certainly, the mature production process has been explored, displaying the highest level of standardization, relatively low manufacturing costs (especially in large-scale production), and a high-yield rate. Based on the traits above, good consistency could be maintained,

where only slight differences could be noted in the performance of different monomers. And, the thermal stability is relatively good: the metal casing is conducive to heat dissipation, and the internal winding structure is relatively controllable in the case of thermal runaway (through pressure relief valve).[9] And, the safety design maturity could be noted, especially the high integration of safety devices such as pressure relief valves with good reliability. In addition, it is easy to automate grouping, mainly ascribed that cylindrical shape is regular, making it convenient to fix in the battery pack using brackets or adhesives.

However, the low space utilization (low grouping efficiency) could be noted, where the unavoidable gaps between cylinders result in low total energy (Wh) in limited space. Moreover, monomer energy density is limited, where the metal casing itself has weight and volume, which is beneficial for reducing the overall weight and volume energy density. And, thermal management is still challenged. Especially, in large battery packs, there are many individual cells, and managing temperature uniformity between cells is quite complex.[10] In addition, the high-rate performance is still limited. In the center of the winding structure, the long heat dissipation paths may result in inferior high-rate charge-discharge performance. Although newer designs such as tab-less/full-tab technology have been greatly improved, their energy storage still needs to be further improved.

2.1.2.2 Prismatic Cell/Square Cell

1. **Typical models and evolution**
   Samsung SDI served as a key driver in 2008, owing to the electrification transition of the BMW company, they successfully supplied prismatic batteries for BMW at first. In 2013, Samsung SDI established a long-term partnership with BMW, entering into the supply chain of German automakers, meanwhile solidifying the global market position of prismatic batteries. In recent years, technological innovation has been further explored. Especially the "Blade Battery" of BYD, the widely adopted CTP (Cell-to-Pack) technology was applied in the industry, accompanied by the enhanced energy density and integration efficiency of battery packs, contributing to strengthening the important position of prismatic batteries in the field of EVs. Nowadays, they have been successfully applied in EVs (the mainstream choice of most automobile enterprises in China, Europe, and South Korea), electric buses, energy storage power stations (especially large container energy storage), and high-end consumer electronics products.
2. **About structural characteristics**
   Metal materials (mainly aluminum, occasionally steel) are always used to form square or rectangular casing through stamping or stretching processes. And, the internal electrode groups can be wound or laminated (Z-fold stacking) into a flat shape into the casing. Stacking is becoming more and more popular, especially in the pursuit of high energy density and low internal resistance scenarios.[12] Herein, the cathodes and anodes are typically connected to corresponding terminals on the top cover of the casing through tabs welded on the electrode assembly. The top cover is an independent component, usually made by combining metal (aluminum) and plastic insulating parts. And, the top cover is highly integrated and contains poles (for connection to external circuits), explosion-proof valves/

relief valves (safety devices that rupture to release gases in case of overpressure), filling holes (sealed), and possibly flip discs (an overpressure or overcurrent protection device). The casing and top cover are sealed with high-strength materials using laser welding (aluminum casing).

3. **About manufacturing process characteristics**
   The shell and lid require precise stamping/stretching and welding. The manufacturing of electrode assemblies (winding or stacking) is crucial. Although the stacking process is less efficient compared to winding, technological advancements are rapid.[13] Liquid injection, sealing (laser welding), chemical aging, and other processes are also necessary. The specification size is more diversified than that of the cylinder, and the standardization degree is not as high as that of the cylinder, but it is also gradually forming series (such as the VDA standard size).
4. **Advantages and disadvantage**
   For prismatic cell/square cell, the high space utilization (high grouping efficiency) could be noted, where the square shape can be closely arranged, which is beneficial for high-effective use of space in the battery pack. And, volume energy density is usually better than cylindrical battery pack. They showed higher structural strength, especially since that metal casing provides good mechanical protection (better crush resistance than soft wrap, but overall strength may be slightly inferior to thick cylinder).[14] Note that, their considerable design flexibility could be found, where size specifications could be customized according to customer needs and wide adaptability. Capacity can be made very large (such as for buses and energy storage). Moreover, owing to the larger heat dissipation surface, the larger plane is conducive to the layout of the heat dissipation structure (such as liquid cooling plate). About their safety, the mature relief valve design, and shell strength contribute to inhibiting rapid internal pressure rise, but perhaps leading to local high-pressure points.

   But, it should not be neglected that the production process is relatively complex, where the requirements for shell manufacturing, top cover integration, welding, and sealing are high.[15] Therefore, the cost was relatively high, in comparison to cylinder batteries, especially for small series customization. Of course, the control of consistency is required to be improved, due to the variety of size specifications and relatively complex structures of monomer batteries. The weight of the metal casing is still a burden (but usually lighter than a cylindrical package of the same capacity). Under long-term cycle or temperature change, the square shell may generate internal stress due to internal gas generation or electrode expansion, which affects life (this should be considered during design).

2.1.2.3 Pouch Cell/Lithium-ion Polymer Cell

1. **Typical models and evolution**
   The investigation of pouch cells was carried out in 1994. Firstly, the initial automotive applications began in 2000. Soon afterwards, with the assistance of AESC, automotive-grade large pouch cells were successfully explored in 2007. And, LG Chem's mass production of PHEV pouch cells began in 2010, Nissan Leaf became the first mass-produced EV equipped with pouch cells. Subsequently, pouch cells were popular among European and American automakers due to

their high energy density. But, their market share was still limited in China. And, their recent advancements were obvious, including industry solutions like SPS addressing assembly efficiency challenges, ongoing localization of aluminum-plastic film production, and the recognition of the pouch structure's advantages as an ideal pathway toward solid-state batteries. Nowadays, they have been applied in numerous fields, such as smartphones, tablets, ultra-thin laptops, wearable devices (smartwatches, headphones), drones, lightweight EVs with high weight and space requirements, and some high-end EVs.

2. **About structural feature**
   Clearly, the core feature is that there is no hard metal shell. And, the internal electrode group is laminated (and rarely wound) into thin sheets. An aluminum-plastic composite film is used as the encapsulation material. The film usually consists of three layers.[16] About the outer layer, the nylon layer (protection and scratch resistance) could be used. About the intermediate layer, the aluminum foil layer served as a barrier to moisture and oxygen, providing structural strength. About the inner layer, the heat sealing layer was selected as modified PP, etc., which can be melted and sealed after heating, and are resistant to electrolyte corrosion. Moreover, the cathodes and anodes are connected through welding at the tabs on the electrode assembly, leading out from the reserved sealed edge of the aluminum-plastic film. And, the stacked electrode group is placed into the expanded aluminum-plastic film, electrolyte is injected, and then the edge of the aluminum-plastic film is heat sealed in a vacuum or inert atmosphere to form a closed "bag." The quality of heat sealing is critical to battery life and safety.[17] There is usually no integrated rigid pressure relief valve. The aluminum-plastic membrane itself bulges or even ruptures when the internal pressure is too high (usually designed with weak points) to relieve pressure.
3. **Manufacturing process feature**
   Aluminum-plastic film would be firstly punching and cutting. Certainly, the lamination process is key (efficiency improvement is the industry focus). Note that liquid injection and vacuum heat-sealing process requirements are extremely high (directly affecting the tightness and yield). After formation aging, additional fixtures for formation or final pressure shaping procedures are often required to improve interface contact, reduce gas, and make the shape smoother. The manufacturing process needs to strictly control the environment (drying rooms) to prevent moisture intrusion.
4. **Advantages and disadvantage**
   The lightest weight and the highest volumetric energy density could be noted. The heavy metal shell is removed, and the aluminum-plastic film is very thin, so the weight energy density and volumetric energy density of the monomer are usually the highest among the three structures. The grouping efficiency is also high (flexible stacking). Meanwhile, the shape, size, and thickness can be highly customized to meet the needs of a variety of shaped spaces, such as ultrathin mobile phones, wearable devices, and car chassis spaces of specific shapes. And, the laminated structure was combined with a short current path (tab can be multi-position design), conducive to reduce internal resistance, improve rate performance, and

reduce heat generation. The safety failure mode is relatively "mild": when the internal pressure rises, the aluminum-plastic film will swell first, providing a certain warning space; the final rupture usually releases less energy than a metal-clad battery explosion (but it is not absolutely safe and may still catch fire). In case of thermal runaway, gases are relatively easy to escape from the sealing edge.

The mechanical strength is the worst. The impact resistance, extrusion resistance, and puncture resistance of aluminum-plastic film are much lower than those of metal casing. Additional structural support and protection design (module frame, flame-retardant insulation material) are required in battery pack.[18] The manufacturing environment has stringent requirements. It is highly susceptible to moisture, necessitating strict control in dry rooms, and has high production costs. The sealing challenge still existed. The heat-sealing process is complex, and there is a risk of electrolyte leakage or moisture intrusion under long-term use (especially in high-temperature and vibration environments), affecting life and safety. The seal at the tab is the weak link. Swelling (gassing) is a more significant problem. Gases generated during charging and discharging are more likely to cause battery swelling, affecting life and stability in modules. Good venting design or electrolyte/additives to inhibit gassing are required.[19] The monomer consistency control is difficult. The high degree of customization and complex manufacturing process (especially lamination and heat sealing) put forward higher requirements for consistency control. The cost of grouping may be increasing. Although the cell is lightweight, the module structure is usually more complex (requiring rigid brackets, end plates, insulating materials, etc.) to protect the fragile pouch cell, which may offset some of the advantages.

A comparison of the performance of different battery types is provided in Table 2.1.

Table 2.1 Comparison of various types of batteries.

Feature	Cylindrical	Prismatic	Pouch
Housing material	Steel shell/ aluminum shell	Steel shell/ aluminum shell (mainstream)	Aluminum-plastic composite film
Internal structure	Jelly roll	Jelly roll or stack	Stack (mainstream)
Mechanical strength	Highest	High	Lowest
Weight energy density (per unit)	Low	Medium	High
Volume energy density (per unit)	Low	Medium	High
Grouped space utilization rate	Low	High	High
Design flexibility	Low	Medium	High

(Continued)

Table 2.1 (Continued)

Feature	Cylindrical	Prismatic	Pouch
Process maturity level	High	High	Medium
Cost (per unit)	Low	Medium	High
Consistency	High	Medium	Medium
Heat dissipation	Long heat dissipation paths for individual modules, complex management within the enclosure	Flat surfaces facilitate heat dissipation	A large surface area facilitates heat dissipation, but requires a protective coating
Representative model	18650, 21700, and 4680	Available in various sizes	Highly customized
Primary application areas	EVs, tools, home appliances, and energy storage	Mainstream EVs, buses, and energy storage	Consumer electronics, wearables, drones, light EVs, and select premium EVs

2.1.3 Factors Influencing the Selection of Shape Structure

In practical applications, the choice of the shape structure of the battery is not determined by a single factor but needs to be considered comprehensively. About application scenario requirements, prioritization of energy density (weight/volume) was always required, accompanied by power density, spatial constraints, mechanical reliability, and cost budget. And, about system integration design, the spatial layout of battery pack, the thermal management scheme (air cooling/liquid cooling), and the structural protection capability were deemed as key elements. About supply chain and cost, maturity of different structures, supplier capabilities, and large-scale production costs were required. About the trends in technological development, the innovations such as seamless tabs on cylindrical cells, CTP/CTC (Cell-to-Chassis) technology, and blade batteries (long, thin prismatic pouch concept) continue to merge and blur the boundaries of traditional structural designs, enhancing system-level performance.

Cylindrical, prismatic, and pouch cells are the three main types of lithium-ion batteries divided according to their shape and structure, each with its distinct structural characteristics, manufacturing process, performance advantages, and limitations.[17] The cylindrical shape is known for its robustness, standardization, and low cost; the prismatic shape strikes a good balance between space utilization and strength; and the pouch cell provides the highest single energy density and extreme design flexibility.[20] Understanding their core differences is the basis for battery selection, system design, and technology roadmap planning. With the advancement of materials science and manufacturing technology, as well as innovation in system integration technology (such as CTP and CTC), various shapes and structures are constantly evolving, learning from each other's advantages and jointly promoting lithium-ion battery technology to higher performance, safer, and lower cost.

2.1.4 Demands Analysis of Battery Assembly Process for Spent Battery Recycling

1. **Standardization and modular design: improving recycling compatibility**
   At present, spent batteries have different shapes (cylinder, square, and pouch cell coexist) and various internal connection modes, which leads to low efficiency in disassembly and even manual intervention. To accommodate automated recycling lines, battery design needs to be standardized in the following areas. About unification of size and connector, standardization of battery cell size (such as 32650 and 18650 cylindrical batteries), and module connector interface can greatly improve the versatility of robot disassembly path planning.[21] About modular packaging technology, the use of snap-in or plug-in module structures that can be quickly separated, avoiding the difficulty of disassembly caused by the curing of traditional potting compounds. About material sorting friendliness, the outer casing material (such as steel casing and aluminum casing) needs to have a clear physical or magnetic difference from the internal metal components (copper foil and aluminum foil) to facilitate eddy current separation.
2. **Material selection and chemical compatibility: reduce recycling difficulty**
   The difficulty of material separation in the recycling process is closely related to the material selection in the battery preparation stage, and the following aspects need to be optimized: binder environmental protection replacement, current collector purity control, and electrolyte formulation optimization.[22]
3. **Easy disassembly structure design: assist physical separation**
   The physical structure of the internal components of the battery directly affects the separation efficiency of the recycling process. The following concepts should be incorporated into the design: simplification of connection mode, identification and traceability, and design for the ease of opening the casing.
4. **Environmental protection and safety design: ensuring the safety of the recycling process**
   The safety risks of spent batteries in the recycling process require pre-prevention and control at the design stage: harmless material replacement, thermal stability enhancement, and leakage protection design.

In summary, the efficient recycling of spent lithium batteries is not an independent link, but a key driving force for battery manufacturing technology innovation. From standardized structure, environmentally friendly material selection, and easy disassembly design to second-life use compatibility, every step of battery preparation and assembly needs to be embedded with the concept of **"Design for Recycling."**[23] This will not only promote the recovery rate and reduce pollution, but also promote the new energy industry to build a true **"closed loop of resources"** and realize the greening of the whole life cycle. With the promotion of China's "two new policies" to replace the old with the new, battery products integrating a recycling-friendly design will occupy the core advantage in the future market competition.

2.2 Detection and Evaluation

The detection of spent lithium-ion batteries is a key step to ensure their safety and performance, mainly including various methods such as physical testing, chemical testing, and electrochemical testing. These detection methods possess intrinsic advantages and disadvantages, suitable for different application scenarios and needs.

2.2.1 Physical Testing Methods

Common physical testing methods are shown in Figure 2.3. Physical testing is the most basic detection method, mainly through appearance inspection, size measurement, and weight analysis to evaluate the overall condition of the battery. Visual detection can identify damage, deformation, and corrosion of the battery casing; size measurement and gravimetric analysis help to determine the wear and changes of materials inside the battery. The advantages of physical testing are simple operation, low cost, and suitable for large-scale preliminary screening. Common physical testing techniques include industrial computed tomography (CT) scanning ultrasonic testing (UT), and infrared thermal imaging.

1. **Computed tomography**
   Industrial CT, as a nondestructive detection technology, has its core principle based on the physical characteristics of X-rays and material interaction and the mathematical reconstruction algorithm.[24] This technology penetrates the tested cell by emitting high-energy X-rays. Because the mass absorption coefficient of different materials (such as metal electrode, diaphragm, and electrolyte) is significantly different, the ray intensity after penetration presents gradient attenuation, and then a projection signal with gray difference is formed at the detector end.[25] Subsequently, the precision rotating stage drives the battery to perform multi-angle scanning, and the high-sensitivity detector array continuously collects two-dimensional (2D) projection data from various angles. Finally, through mathematical methods such as filtered back projection (FBP) and iterative reconstruction (such as the ART algorithm), mass projection data are converted into cross-section tomographic images, and construct a submicron resolution 3D model of the internal structure through 3D spatial registration and voxel fusion technology, which can clearly display key defect characteristics such as pole piece cracks, lithium dendrite growth, and electrolyte distribution.
2. **Ultrasonic testing**
   UT is a nondestructive testing technique that uses the characteristics of high-frequency sound waves (usually 0.5–20 MHz) propagating in materials to detect defects or structural abnormalities inside objects by analyzing signal changes of reflected waves or penetrating waves.[26] The core process includes exciting high-frequency sound waves through piezoelectric probes and injecting them into the tested battery. The injected sound waves are reflected during propagation due to contact material interfaces (such as electrode coating peeling, electrode

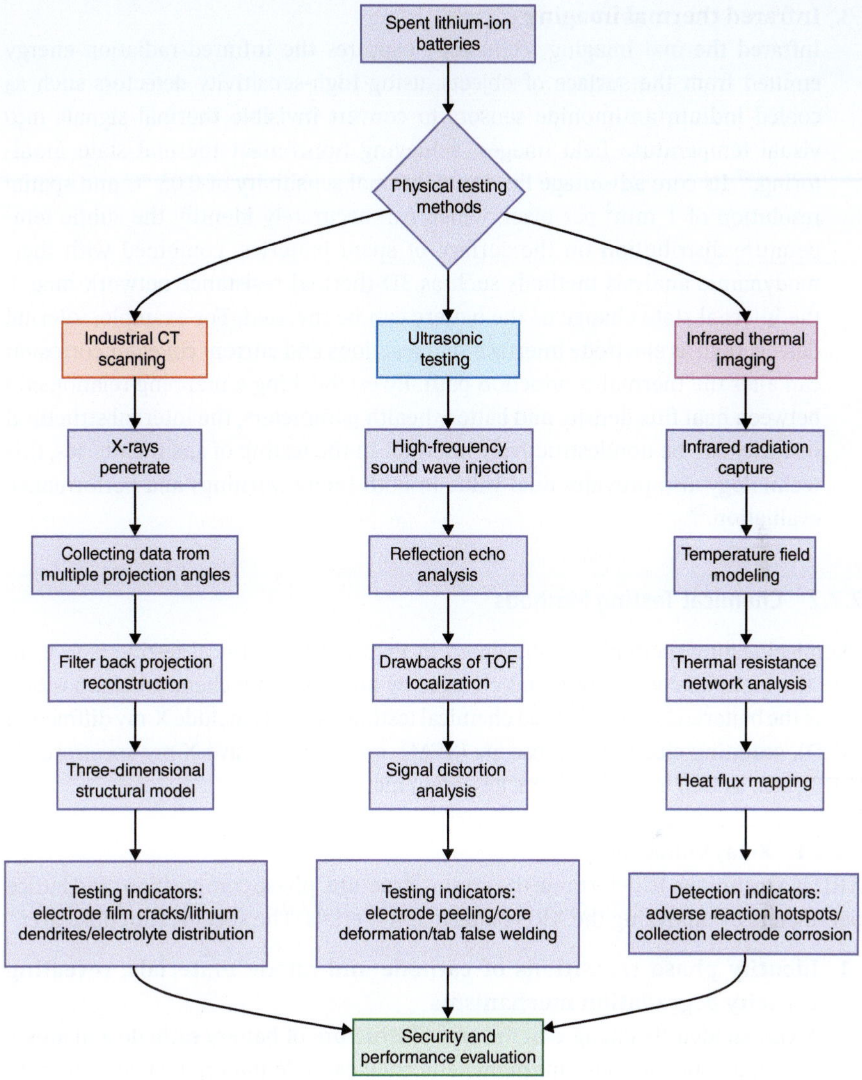

Figure 2.3 Physical testing methods.

plate cracking, or diaphragm perforation). At this time, the probe synchronously receives the reflected echo signal, locates the defect depth according to the echo time difference (TOF), and analyzes the defect size and type (such as porosity and delamination) in combination with signal intensity and waveform distortion (such as phase inversion and amplitude attenuation). In the detection of spent lithium-ion batteries, this technology can accurately identify structural abnormalities such as electrode active material peeling, core mechanical deformation, and false welding of tab terminals. Simultaneously assess the distribution state of the electrolyte indirectly through the speed-density correlation model, providing key data support for safety evaluation in the second-life use.

3. **Infrared thermal imaging**
   Infrared thermal imaging technology captures the infrared radiation energy emitted from the surface of objects, using high-sensitivity detectors such as cooled indium antimonide sensors to convert invisible thermal signals into visual temperature field images, achieving noncontact thermal state monitoring.[27] Its core advantage lies in its thermal sensitivity of 0.05 °C and spatial resolution of 1 mm^2 per pixel, which can accurately identify the subtle temperature distribution on the surface of spent batteries, combined with thermodynamic analysis methods such as 3D thermal resistance network model, the internal state change of the battery can be inverted. For example, internal defects such as electrode interface side reactions and current collector corrosion can alter the thermal conduction path. By establishing a mapping relationship between heat flux density and battery health parameters, the internal structural integrity can be nondestructively inferred. In the testing of spent batteries, this technology also provides dual value in both safety warnings and performance evaluation.[28]

2.2.2 Chemical Testing Methods

Chemical testing methods are presented in Figure 2.4. Chemical testing is to evaluate the material composition and changes by analyzing the chemical composition inside the battery. Commonly used chemical testing methods include X-ray diffraction (XRD), scanning electron microscopy (SEM), energy-dispersive X-ray spectroscopy (EDS), etc., as well as in situ characterization methods.

2.2.2.1 X-Ray Diffraction

XRD is a technique to determine the crystal structure, phase composition, and lattice parameters by analyzing the XRD pattern of materials. The specific functions are:

1. **Identify phase transitions of cathode and anode materials, revealing capacity degradation mechanisms**
   XRD can identify changes in the crystal structure of battery cathode and anode materials, such as determining whether the cathode material (such as ternary

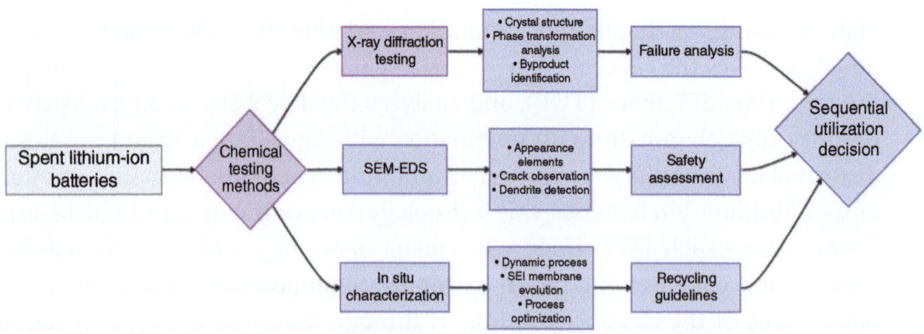

Figure 2.4 Chemical testing methods.

materials: nickel cobalt manganese oxide [NCM] and LFP) has degraded, measure changes in cell parameters (such as a-, c-axis length), and evaluate the degree of material expansion/contraction to determine lattice parameter changes.[29] For anode materials, the change of interlayer spacing of the graphite anode can be detected to detect whether the graphite structure is degraded. For silicon-based anodes, one can detect whether silicon undergoes a crystalline-to-amorphous phase transition during charging and discharging to evaluate if a phase change has occurred in the silicon-based anode.

2. **Detect electrolyte decomposition products, evaluate chemical stability**
   XRD can analyze and detect electrolytes and by-products, such as LiF generated from the decomposition of lithium hexafluorophosphate ($LiPF_6$) or other by-products like $LiPO_2F_2$.
3. **Identify impurities and contaminants, optimize recycling processes**
   XRD can identify copper current collector corrosion products such as Cu^+/Cu^{2+} oxides (like CuO and Cu_2O) or residues of metals like nickel and cobalt. In addition, for detecting foreign substances mixed in the disassembly process (such as Fe particles and Al chips), can also locate its phase through characteristic diffraction peaks.

2.2.2.2 Scanning Electron Microscopy

SEM and EDS can observe the microstructure and elemental distribution of materials. SEM uses a focused electron beam to scan the sample surface, generating high-resolution (nanometer-level) microstructure images that reveal the surface structural characteristics of the material. EDS is combined with SEM to analyze the elemental composition and distribution of materials by detecting the characteristic X-rays emitted by samples excited by electron beams.[30] For cathode materials (such as NCM and LFP), particle crack, pulverization, and agglomeration can be observed by SEM and EDS. Anode materials (such as graphite, silicon-based) can detect graphite delamination, silicon particle fragmentation, or lithium dendrite growth. In addition, it can also check the pore clogging, tearing, or thermal shrinkage of the separator to evaluate its safety performance.[31]

2.2.2.3 In Situ Characterization Technique

In the detection and second-life use of spent lithium-ion batteries, in situ characterization technology provides a unique perspective on the microscopic scale for accurate evaluation of aging mechanisms and residual performance by dynamically monitoring the coupling evolution of internal materials and electrochemical behavior of batteries in real time. This technology can directly observe the structural phase transition of electrode materials (such as the transition from cathode layered structure to spinel structure), lithium-ion insertion/extraction kinetics, and volume expansion effect during charge and discharge, synchronously correlate the changes of macroscopic parameters such as voltage, current, and temperature, so as to analyze the dominant factors of capacity decay (such as the competition mechanism between cathode active material loss and anode lithium evolution) and identify the critical point of SEI film thickening or cracking, providing key criteria for thermal runaway warning. With the help of high-resolution imaging means (such as in situ SEM/TEM), the evolution law of microscopic defects such as lithium dendrite nucleation growth and polar plate crack propagation can be dynamically captured, and the safety screening accuracy of graded battery packs can be significantly improved. In addition, this technology can also guide the optimization of the recovery process, for example, by in situ XRD monitoring the phase transition kinetics of LFP cathode during acid leaching (such as the conversion rate of $FePO_4{\cdot}2H_2O$ to amorphous phase), determining the optimal leaching time and acid concentration, and improving the metal recovery rate.

The advantage of chemical testing is that it can provide detailed material information and help to understand the failure mechanism of batteries. However, chemical testing often requires complex sample preparation and expensive instrumentation, which is costly and time-consuming. And chemical testing cannot directly reflect the battery capacity, internal resistance, cycle life, and other key performance indicators, but should be combined with charge and discharge tests or EIS and other electrochemical means of comprehensive evaluation.

2.2.3 Electrochemical Testing Methods

Electrochemical testing methods are depicted in Figure 2.5. Electrochemical testing is to evaluate the key parameters such as capacity, internal resistance, and cycle life by measuring the electrochemical performance of batteries. Common electrochemical testing methods include cyclic voltammetry (CV), electrochemical impedance spectroscopy (EIS), and constant current charge and discharge testing, among others.

1. **Cyclic voltammetry**
   CV is an electrochemical testing method that involves applying a linearly varying voltage to an electrode and measuring the current response to analyze the material's redox properties, reaction kinetics, and electrode stability. Its core principle involves exciting electrode reactions through voltage scanning, recording current-voltage curves (CV curves), and thereby revealing the electrochemical behavior of battery materials.[32] By comparing the redox peak areas,

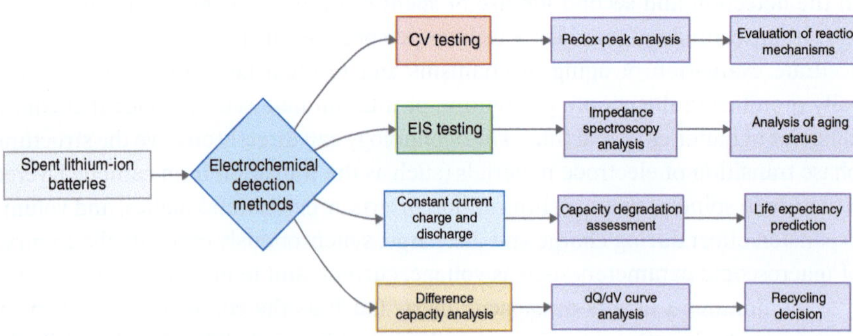

Figure 2.5 Electrochemical detection methods.

the degree of capacity decay can be accurately quantified. At the same time, based on the peak potential difference and peak symmetry, the reversibility of the reaction can be judged, and batteries with excellent electrochemical stability can be selected for second-life use. The shift of the CV curve characteristic peak can reveal the aging mechanism of cathode material phase transformation, anode lithium evolution, and SEI film thickening, which provides a basis for directional repair or disassembly recovery.

2. **Electrochemical impedance spectroscopy**
   EIS is a technique for analyzing the electrochemical process inside a battery by applying small sinusoidal alternating current signals of different frequencies (usually in the frequency range of 0.01 Hz–100 kHz) to the battery and measuring its impedance response.[33] Its core principle is to model the battery as an equivalent circuit (such as a Randles equivalent circuit), and by fitting impedance spectroscopy data, it analyzes the changes in each component (such as ohmic resistance, charge transfer resistance, diffusion resistance, etc.), revealing the mechanisms of battery aging. In the detection of spent lithium-ion batteries, similar to CV testing, EIS can be used to assess battery State of Health (SOH), as the increase in charge transfer resistance is associated with a decrease in electrode activity (e.g., phase transition of cathode materials, thickening of cathode SEI films), and capacity decay is also reflected.[34] Ohmic impedance can be used to analyze the internal resistance, and when the ohmic impedance increases, it indicates that the current collector may be corroded or the connection interface may be deteriorated. At the same time, EIS can also analyze the aging mechanism of the battery, detect the growth of SEI film, the decomposition of electrolyte, and the occurrence of side reactions. For tiered utilization screening, EIS can compare the impedance spectra of multiple batteries, selecting modules with R_{ct} differences less than 20% for energy storage systems to ensure balanced performance of the battery pack. It can also investigate safety risks by detecting abnormal impedance arcs, indicating micro short circuits or lithium dendrite growth, and eliminating safety hazards.
3. **Constant current charge and discharge testing**
   Constant current charge and discharge testing is a common method to evaluate battery capacity and cycle life. The technology simulates the actual charging and discharging behavior of the battery, and obtains key parameters (such as capacity, internal resistance, voltage characteristics, etc.) through multiple charging and discharging cycles of the battery, thus providing a scientific basis for the second-life use or disassembly of the battery. The residual capacity reflects the ratio of actual discharge capacity to rated capacity to judge whether it is suitable for second-life use. The change of internal resistance is calculated by voltage plateau change or pulse test to reflect the aging degree of the battery. Abnormal voltage plateau may indicate internal short circuits, lithium precipitation, and other problems. Constant current charge-discharge tests can quickly identify rechargeable batteries, extend their life cycles, and detect potential failures (such as internal short circuits and thermal runaway risks) to ensure subsequent processing safety.

4. **Differential capacity analysis**
   Differential capacity analysis (dQ/dV analysis) is a refined diagnostic technique based on charge/discharge curves for spent lithium-ion battery recycling. It reveals the internal aging mechanisms and failure modes of the battery by analyzing the differential characteristics of battery capacity with respect to voltage changes, providing deeper insights for evaluating the SOH of the battery and making recycling decisions. DQ/dV analysis enables accurate recovery and avoids blind disassembly. Design recycling paths for different aging types (e.g., priority extraction of metals from anode failure batteries, priority harmless treatment of electrolyte leakage batteries). It can also help realize the safety of the second-life use, eliminate high-risk batteries through lithium detection, and reduce the fire hazards of energy storage systems. In addition, dQ/dV is good for supporting circular economy, which can promote battery recycling from "extensive disassembly" to "fine diagnosis-directional regeneration" upgrade.[35]

In summary, the detection methods of spent lithium-ion batteries have their own characteristics and functions, which are suitable for different application scenarios and requirements. Physical testing is suitable for initial screening and rapid assessment; chemical testing can provide detailed material information suitable for in-depth analysis; electrochemical testing can directly reflect the performance and state of the battery and is suitable for accurate evaluation. In practical applications, it is usually necessary to combine multiple detection methods to comprehensively evaluate the safety and performance of spent lithium-ion batteries.

2.2.4 Comparison and Analysis of Detection Technologies

2.2.4.1 Insufficient Detection Accuracy

As the number of cycles increases, the performance of the battery will decline due to physical and chemical changes in the battery, specifically reflecting capacity decay and impedance rise. Moreover, the capacity of the battery is also irreversibly lost due to increasing aging, depending on the self-discharge rate and storage conditions of the battery. Since battery aging is closely related to changes in battery state, and this relationship is always nonlinear, it must be considered an important factor when constructing detection models.

Despite extensive research on the current aging mechanisms of lithium-ion batteries and some significant findings, battery aging remains difficult to identify and quantify. Many factors, such as temperature, storage state of charge (SOC), number of cycles, and depth of charge/discharge, play a direct or indirect role in the aging process. The weights of these factors and their interactions are unclear. The training dataset used to build the predictive model is obtained under monotonic conditions, so all variables cannot be considered. Then the original model already has natural errors and limitations due to ignoring or misinterpreting other factors. The complexity and computational burden of the model make people less interested in exploring multifactor models, and presetting more influential parameters for testing models will undoubtedly improve the accuracy of prediction results. A feasible approach is to control the major factors that significantly impact the aging process, while using other factors as bias in the estimation model. However, a deep understanding of battery aging mechanisms is needed to identify the dominant factors and weigh other factors in different aging processes, which remains a challenging issue.

Most of the existing detection technologies focus on judging the health status of batteries.[36] Admittedly, this is a key factor affecting the second-life use of spent lithium-ion batteries. However, as an indicator of battery state, SOH does not exist in isolation. Traditional battery state description parameters also include battery SOC and remaining useful life (RUL), among others. These states are not independent; in practice has a complex coupling relationship with each other. For SOC detection, capacity and resistance are often considered critical parameters. Meanwhile, SOH degradation of batteries can be reflected by capacity loss or resistance increase, and accurate SOH data is the basis and premise of RUL prediction.[37] Therefore, it is significant and necessary to propose some methods to realize multistate joint estimation. In the existing estimation methods, singlet estimation methods account for the majority. These methods have high accuracy in laboratory settings, but are not ideal for practical applications. Collaborative estimation methods tend to obtain more reliable prediction results. One method of collaborative estimation is to select special parameters related to multiple battery states to build a prediction model, and use the results of one state to update the model. Although this method improves the accuracy of the results, the difficulty and complexity of model construction become a new problem. Moreover, the difficulty of modeling increases with the number of estimated states, so many studies focus on the joint estimation of two states. New thinking is correcting the predicted result of one state by introducing other state estimation results as deviations. This method can reduce the computational burden of the model construction phase. It is worth noting, however, that false corrections can occur without a deep understanding of the coupling between different states.

In addition, although some current studies have deeply analyzed the degradation mechanism of battery materials by means of various physical and chemical characterization and in situ characterization methods, and have clarified some characteristic parameters highly related to battery performance, obtaining these parameters often depends on high-end detection equipment and a large number of destructive steps. This makes it meaningless to work on large-scale battery testing. Therefore, thoroughly analyzing the battery aging mechanism and finding easily measurable characteristic parameters that have a high correlation with battery aging is crucial for achieving efficient and accurate battery detection technology.

2.2.4.2 Low Detection Efficiency

Due to the variety and large number of spent lithium-ion batteries, detection technology should pay more attention to the speed and efficiency of detection. Therefore, large-scale battery state estimation becomes a new challenge. The difficulty of large-scale battery state estimation lies in selecting appropriate characteristic parameters to construct detection model.[38] More representative parameters are always harder to obtain, which also means that the estimation is inefficient. However, easily obtainable parameters such as voltage, resistance, or temperature cannot guarantee the accuracy of predictive models. For large-scale battery estimation, the selection of characteristic parameters needs to strike a balance between prediction speed and accuracy.

Most of today's spent lithium-ion batteries have three levels of structure: battery cells, modules, and battery packs. Therefore, one of the reasons for the low efficiency of echelon utilization detection is that it is not clear which level to select for detection and recombination, which directly affects the technical difficulty and related cost. The advantages of direct detection and the second-life use of battery pack specifications lie in low consumption, no need for complex and low disassembly steps, strong economy, and high efficiency. However, due to the difference in operating conditions, although the batteries in the same battery pack still have obvious consistency differences, their consistency is obviously reduced after further processing, and it is easy to cause safety problems during use. The operation of detection and hierarchical utilization at the module level is relatively simple and highly flexible, but in addition to the consistency problem that cannot be ignored, a single module is likely to cause the entire module to be unable to be utilized due to partial battery damage, resulting in additional economic losses. On the other hand, the current detection technology still has a long detection speed, often reaching the minute level, which is difficult to meet the detection needs of large-scale decommissioned lithium-ion batteries.[39]

At present, most of the detection and second-life use work revolves around the single-cell battery, which has the advantages of maximizing the reuse of the residual value of the spent battery, high battery utilization rate, and high consistency of the battery after reorganization, and is not easy to cause safety problems. However, the defect is that the disassembly process will bring high economic consumption, and the battery is easily damaged during the disassembly process. And the power battery model is diverse, the structure is complex (such as different packaging forms and adhesive use), and disassembly needs physical and chemical methods combined, it is easy to cause electrolyte leakage or short circuit risk. At the same time, the investment in dismantling equipment is large, and the economy is lower than that of direct scrapping. In addition, the detection work requires battery operation one by one, which is difficult to meet the rapid sorting requirements of large-scale spent batteries. It cannot adapt to the pipelined detection scenario, which becomes the efficiency bottleneck of the recycling production line.

2.2.4.3 Insufficient Historical Data

The data-driven detection method has unique advantages; that is, it has flexibility and model-free characteristics when establishing a state prediction model. After successful modeling, the detection rate is fast and the accuracy is high. Therefore, it is regarded as the key to realize the rapid detection of spent lithium-ion batteries. However, similar to the dilemma of traditional machine learning methods, the accuracy of training data seriously affects the application of such detection methods. Data-driven methods build predictive models by measuring the relationship between input signals and changes in battery state, while ignoring electrochemical reactions in the battery. Therefore, the effectiveness of these methods depends largely on the validity and volume of data. Unfortunately, it is difficult to obtain high-quality data due to the long time required for battery testing and the complexity and variability of its operating environment. Rich historical data can help improve prediction accuracy.

However, historical data on spent lithium-ion batteries are mostly missing or the sampling frequency is insufficient. In this case, only test data can be used to establish a test model for spent lithium-ion batteries. In order to improve prediction accuracy, more test data is needed, which leads to longer test time, which is unrealistic for large-scale life prediction of spent lithium-ion batteries.

During the operation of the power battery, the battery management system will record a large number of battery operating conditions, and the accuracy of the training model can be greatly improved by borrowing such historical data. However, such data is generally controlled by cloud systems of different enterprises and is usually not open to the public, making it difficult to obtain. In the application of machine learning technology in the detection and the second-life use of spent lithium-ion batteries, data quality and acquisition problems constitute basic bottlenecks. The phenomenon of data islands is common in the industry. The data barriers among automobile enterprises, battery factories, and recyclers lead to the training set size of a single machine learning model being less than 100,000 groups (ideal demand exceeds 1 million), which is only one tenth of the critical value of effective training.[40] Taking the FedBatt federal learning system dominated by CATL as an example, although 40 enterprises share 200 TB battery life cycle data, data encryption for commercial confidentiality requirements leads to 15–20% loss rate of key features, which seriously weakens the recognition ability of the model for complex aging modes (such as lithium evolution and SEI thickening coupling effect). In addition, due to the lack of relevant data management system, the storage mode of these data is complex, the data format is not uniform, and the historical traceability is difficult, resulting in its weak availability.[41]

2.2.4.4 Vision for the Development of Detection Technology

1. **Improve the detection accuracy**

   Accurate detection of the battery state of spent lithium-ion batteries is crucial for achieving the second-life use and subsequent recycling processes. However, due to the wide variety of battery materials on the market today, spent lithium-ion batteries exhibit significant material differentiation characteristics. According to the traceability data of spent batteries in 2023 by the Ministry of Industry and Information Technology, LFP accounts for 58.6%, ternary materials (NCM/NCA) account for 37.1%, and LCO and lithium manganese oxide (LMO) account for 4.3% in total. The aging paths dominated by different cathode material systems are essentially different. For example, for LFP batteries, the main decay mechanisms are loss of lithium inventory (LLI) and loss of active material (LAM). The continuous thickening of SEI film leads to significant increase in ohmic impedance, and the capacity decay presents linear characteristics; for ternary battery systems, they typically face complex degradation issues such as the dissolution of transition metals (especially the disproportionation reaction of Mn^{3+}) and the collapse of crystal structures (transition from layered to spinel phase). The increase in internal resistance exhibits a nonlinear abrupt change characteristic; however, the LCO system is mainly limited by structural stability, cobalt dissolution, and particle microcrack propagation easily occur, and the capacity diving phenomenon is obvious. The difference of aging mechanism leads to the problem that the existing detection technology generally has low detection accuracy.[42]

Moreover, existing detection technologies mostly rely on a single feature parameter, such as analyzing the inflection points of the differential voltage (dV/dQ) curve or extracting the relaxation voltage time constant. However, the battery is involved in a variety of electrochemical reaction processes, and its macroscopic physical and chemical parameters are often affected by many factors, including external conditions, which leads to unclear correlation with the battery state and is difficult to capture, resulting in poor detection accuracy. Therefore, it is urgent to develop detection technology with higher detection accuracy to provide technical support for the second-life use of spent batteries.

2. **Improve detection efficiency**

The low detection efficiency in the current field of power battery second use is due to the high heterogeneity of battery products. The market for spent lithium-ion batteries has developed into a multidimensional technological roadmap, featuring various physical forms such as pouch cells (thickness 0.5–8 mm), blade cells (length 960 mm), and cylindrical cells (such as 18650/21700), among others. Their capacities range from 2.6 to 300 Ah, and their voltage platforms are distributed between 3.2 and 3.7 V. This diversity in physical form directly leads to existing detection equipment being difficult to accommodate. More prominently, in order to meet the stringent requirements of new energy vehicles for energy density (current mainstream requirements $\geq$ 180 Wh/kg), modular integrated design is generally adopted for power battery systems. Typical power battery packs usually integrate 96–200 single-cell batteries, forming a multilevel module structure through irreversible assembly processes such as laser welding and structural adhesive bonding. According to industry statistics, dismantling a standard battery pack (containing 100 cells) requires 4–6 man-hours, resulting in an increase in recycling costs of about 1200 yuan/ton. This system-level disassembly not only significantly increases the marginal cost of the second-life use but also causes the number of units to be tested to grow exponentially—from the cell level to the system level, the sample size can increase by 200 to 500 times, directly leading to a decrease in testing efficiency by more than 60%.

In order to meet the demand for large-scale sorting of spent batteries, it is necessary to realize efficient screening on the premise of ensuring the structural integrity of batteries. Under the premise of not destroying the structure and performance of the battery, the chemical composition and structural information inside the battery are obtained to detect defects and material changes inside the battery without causing damage to the battery. The electrode material aging status can be evaluated online without disassembling the battery, balancing detection efficiency and battery reuse value. In addition, for the large-scale sorting of discarded lithium-ion batteries, speed and accuracy are very important. One of the significant challenges in large-scale, second-life use of spent lithium-ion batteries is designing a detection and classification technology that offers excellent sorting speed and accuracy, as well as good dynamic performance for different application scenarios.[43]

2.3 Safe Discharge

The cell disassembled from the spent battery pack usually has a residual amount of electricity. When the cell is broken, the cathode and anode are easy to contact each other, resulting in a short circuit, releasing the chemical energy stored inside, resulting in a transient increase in temperature, causing fire and even explosion.[6] Therefore, in order to avoid accidents, it is usually necessary to carry out discharge treatment before the cell is broken to release the chemical energy stored inside. Similarly, discharging treatment plays a crucial role in the recycling of lithium resources. During the directional discharge process, lithium ions as charge carriers are deintercalated from the graphite anode and migrate to the cathode material through the electrolyte to complete reinsertion. This mechanism ensures that lithium elements return to the cathode lattice structure from the anode, thereby improving the lithium recovery rate during the hydrometallurgical process of subsequent cathode materials.[44]

The methods of battery discharge can be divided into physical discharge methods and chemical discharge methods. Before discharge, it is necessary to evaluate the safety of spent batteries, and select different discharge processes according to the appearance and capacity of spent batteries. For batteries that have intact external structures and complete functions, with no signs of smoking, fire damage, leakage of electricity, liquid leakage, water immersion, etc., it is advisable to choose physical discharge methods, such as the physical load discharge method (Figure 2.6a), for discharging. For batteries that exhibit one or more of the following conditions: damage to appearance, swelling, distortion, smoking, overheating, leakage of electricity, leakage of liquid, water immersion, short circuiting, insulation failure, having undergone destructive testing but with good appearance, inability to connect to discharge equipment, or failure of the external circuit discharge method, it is advisable to choose chemical discharge methods such as the solution soaking discharge method for discharging (Figure 2.6b).[45]

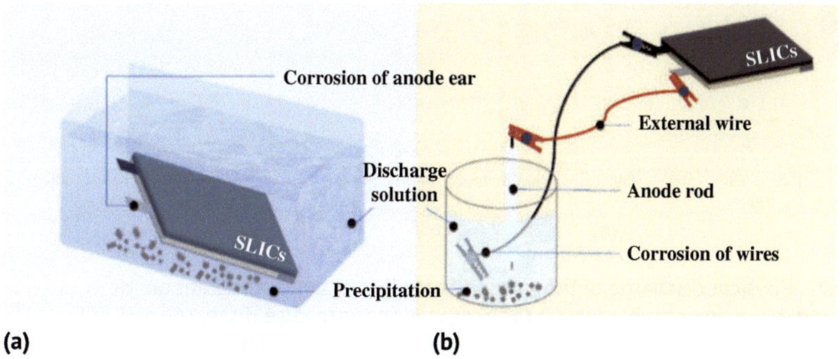

Figure 2.6 Schematic diagrams of discharge for (a) internal immersion method and (b) external solution method. *Source*: [46] / with permission of Elsevier.

2.3.1 Physical Discharge Method

A schematic of the physical discharge process for lithium-ion batteries is shown in Figure 2.7. Physical discharge methods are generally divided into energy-saving feedback discharge and physical load discharge.

1. Energy-saving feedback discharge is a discharge method that uses advanced power electronics technology to convert DC power emitted by battery packs or battery cells into AC power by an active inverter and feedback to the power grid. The method has high safety, can effectively integrate the residual electric quantity of the spent battery, makes the material useful, and maximizes the energy regeneration utilization rate. However, AC/DC inverters are commonly used in existing power systems. When the energy of spent batteries is low or the batteries are over-discharged, the energy feedback efficiency will decrease significantly. In addition, due to the significant inconsistency between batteries, the integration, automatic discharge control complexity, and equipment investment cost are high.[47]
2. Physical load discharge is a discharge method that converts electrical energy into heat and other energy by connecting loads at both ends of the battery. Depending on the type of load, it can be further classified into resistive load discharge method, electronic load discharge method, and conductive powder load discharge method.[45]

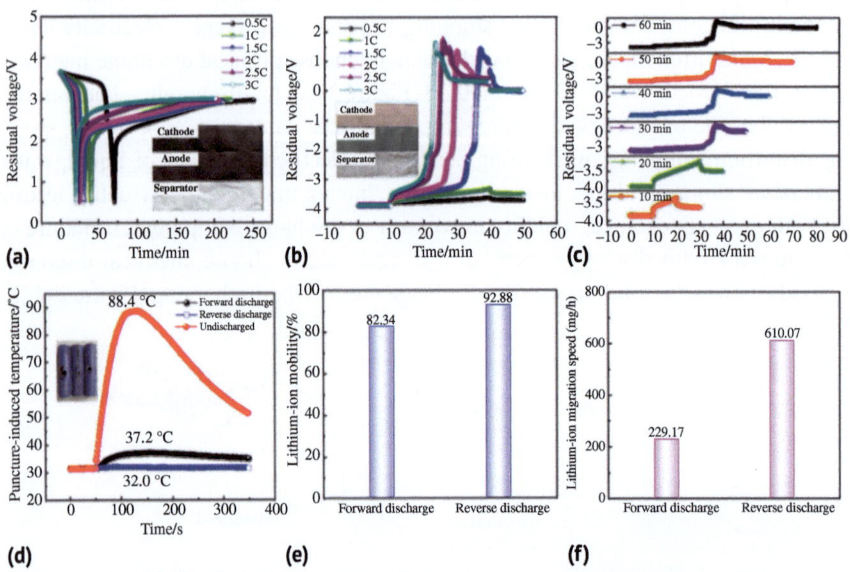

Figure 2.7 Physical discharge of lithium-ion batteries: (a) current regulation for forward and (b) reverse discharges; (c) time regulation for reverse discharge at 1.5 °C; (d) temperature change over time after battery puncture; (e) lithium-ion mobility; (f) migration speed of retired LIBs after forward and reverse discharge at 0.5 and 1.5 °C. *Source*: [47] / with permission of American Chemical Society.

The discharge method using a resistive load involves connecting spent batteries to a resistor to consume battery power through an exothermic process. The discharge of resistive loads can maximize the release of electrical energy, with minimal voltage rebound. However, the thermal radiation generated during the discharge process can cause the surrounding environment temperature to rise, easily leading to fires. In order to conduct heat, load resistors are often placed in cooling bodies fitted with cooling plates. Depending on the discharge equipment, constant current discharge or constant power discharge can also be set during discharge. The cost of the discharge device for constant resistance is relatively low, and it is relatively safer due to limited maximum current and lower power, but the discharge time is usually longer. Constant current discharge devices are more expensive than constant resistance discharge devices, but they have higher discharge efficiency.[48] By adjusting the discharge current using a potentiometer, the discharge power can be manually controlled. The safety level is moderate, but there is still a risk of manual overload due to excessively high current. Resistive load discharge can be applied to large batteries (such as batteries in EVs), but it is difficult to fully apply to small batteries used in electronic equipment.

The electronic load discharge method is to consume the residual energy in the battery by controlling the internal power or the conduction amount of the transistor and relying on the dissipated power of the power tube. Electronic load discharge can count the actual discharge amount in the discharge process, which is suitable for single-cell batteries, but there is a voltage rebound phenomenon after discharge.[49]

The conductive powder short-circuit discharge method involves using highly conductive materials such as graphite and copper powder to short-circuit the battery. The method of discharging conductive powder is simple, the battery discharges quickly, and it does not require the use of complex equipment, making it suitable for deactivating small batteries. However, due to the external conductor discharging easily, causing thermal energy and internal pressure accumulation, the temperature inside the system can quickly rise during discharge, which may then trigger an explosion. Therefore, during the electric powder discharge process, it is necessary to closely monitor heat generation and gas production, equip cooling devices, and ensure that heat and internal pressure are released stably and safely. At the same time, selecting appropriate external conductors and optimizing the discharge rate will significantly enhance the safety of external short-circuit discharges. However, due to the explosion risk of conductive graphite powder itself and the easy oxidation of metal powder surface, this method is difficult to apply on a large scale.

2.3.2 Chemical Discharge Method

A schematic of the chemical discharge process is illustrated in Figure 2.8. The chemical discharge method involves immersing the spent battery in a conductive medium (such as salt solution) to create a short-circuit, thereby achieving complete discharge.[50] During chemical discharge, the energy generated is stored in the solution system, which can avoid the safety risks caused by excessive heat

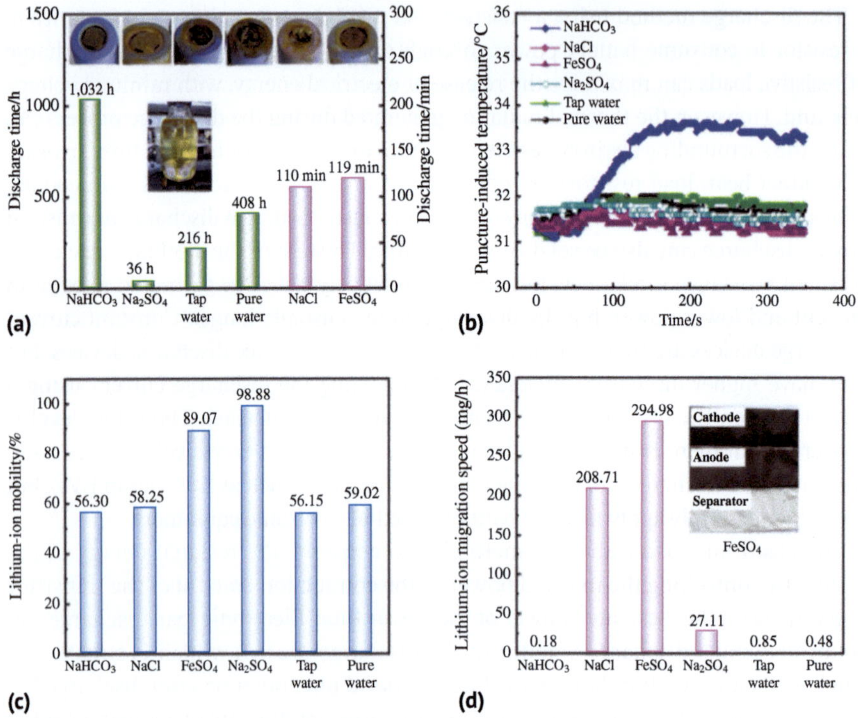

Figure 2.8 Chemical discharge: (a) residual voltage when discharge time reaches 0.5 V; (b) change in temperature over time after battery puncture; (c) lithium-ion mobility; (d) migration speed after discharge. *Source*: [47] / with permission of American Chemical Society.

and gas generation. Therefore, the chemical discharge method is safer than the physical discharge method, which has problems such as high heat release and poor safety. Sodium chloride, as a chemical reagent with good conductivity, low cost, and abundant resources, is widely used in solution discharge. However, the discharge cycle of sodium chloride solution is long; even at the optimal concentration of 10 wt.%, the complete discharge process still takes 18 to 24 hours, which is not conducive to practical production.

The discharge rate of spent batteries in sodium chloride solution can be accelerated by means such as external field reinforcement. For example, using ultrasonic technology or changing the magnetic field strength can effectively increase the migration rate of ions in the solution, enhancing the discharge effect. However, during this process, salt solutions represented by sodium chloride are prone to electrolysis, producing a series of side reactions, such as chlorine gas, hydrogen gas, and oxygen being generated during the discharge of sodium chloride solution. Among them, chlorine gas is a toxic gas that poses potential hazards to the environment.[51] Using a discharge system with a solution that has a lesser impact on the battery casing, such as a sodium sulfate solution system, can effectively reduce side reactions, improve discharge efficiency, and minimize environmental impact. Similarly, adding extra additives such as ascorbic acid to the solution also helps increase its conductivity, accelerate the discharge process, and shorten the discharge time. However, the additive loss is large in the discharge process, and large-scale use will increase the discharge cost. Therefore, selecting low-cost, high-stability, recyclable additives, reducing discharge costs, and improving industrial scale are the key directions for the future.[52]

Moreover, in the field of non-neutral medium discharge, acid/base solution systems can accelerate the reconstruction of the electrical double layer through the high mobility of H^+, or reduce the solution impedance by relying on the high conductivity of OH^-, thus exhibiting unique kinetic advantages. However, because lithium-ion battery casings are typically made of aluminum or steel, they are prone to corrosion when discharged using acidic or alkaline solutions, which can lead to issues such as casing rupture and electrolyte leakage. In severe cases, toxic hydrogen fluoride (HF) gas may be released, posing risks to environmental safety.[53] At the same time, wastewater generated from acid/base solution discharge systems usually contains heavy metals and causes significant pollution. Using low-toxicity, bio-based organic weak acids, such as citric acid and malic acid, will to some extent alleviate the aforementioned issues.

Among the above discharge methods, load resistance discharge and solution immersion discharge are more commonly used. On one hand, discharging load resistors can effectively integrate the energy from spent batteries, coupling it with power-consuming industries such as electrolytic hydrogen production, electrolytic aluminum production, and electrocatalytic ammonia synthesis, thus maximizing resource utilization. On the other hand, solution immersion discharge has better controllability, lower costs, simpler operation, and higher stability for large-scale production.[54]

- **Safety discharge technology analysis and shortcomings:** Before breaking down the battery, it is usually necessary to perform a discharge treatment on the battery. Currently, battery discharge mainly occurs through solution discharge. If sodium chloride solution is used for discharge, the process is slow, incomplete, and exhibits voltage rebound phenomena.[47] At the same time, the discharge of sodium chloride solution is easy to induce electrolytic reaction and generate toxic and harmful gases such as chlorine. When chlorine dissolves in water, it forms a hydrochloric acid solution, which is highly acidic. Due to the battery casing being mostly made of steel or aluminum, the generated hydrochloric acid will strongly corrode the battery casing, causing it to break and leading to significant leakage of internal components, especially the electrolyte.[55] On one hand, the electrolyte contains a large amount of organic solvent components, which easily pollute the environment and trigger fires. On the other hand, the $LiPF_6$ component in the electrolyte forms LiF and PF_6 when dissolved in water, simultaneously releasing HF gas. The strong corrosiveness of HF gas further exacerbates the corrosion of the battery casing, inducing further leakage of battery components. Moreover, during this process, the occurrence of side reactions such as electrolysis further reduces the depth of discharge of the battery, making it impossible to fully release the residual power of the battery, which poses potential safety hazards for subsequent crushing.

- **Development requirements of safe discharge technology:** Establishing an efficient and safe residual energy release and multi-path utilization system.

Spent battery packs still retain some residual power. If they are not properly handled before disassembly, there are significant safety risks such as thermal runaway, short circuits, and explosions. In addition, if the residual electricity is not fully released, it will cause waste of resources. Therefore, it is urgently necessary to establish a green, efficient, and intelligent deep discharge system, while exploring multiple energy recovery pathways to achieve the resource utilization of residual energy.[50]

The current main challenges include uneven residual energy distribution, inaccurate detection algorithms, insufficient safety control during the discharge process, and limited paths for utilizing residual energy. Therefore, it is necessary to optimize the system from three dimensions: detection, discharge, and transformation. The specific directions are as follows:

- **Residual battery capacity online detection and evaluation system:** Construct a detection algorithm based on the voltage-capacity model and thermoelectric response data fusion, achieving precise calculation and tiered classification of individual cell residual capacity.[55]
- **Safety automatic discharge control system:** Develop an intelligent control platform supporting multi-cell parallel discharge, integrate temperature control and state feedback mechanism, ensure "zero spark, zero emission" in the discharge process, and achieve discharge depth of 100%.[56]
- **Multi-energy conversion platform:** Design and integrate thermoelectric/mechanical energy conversion modules, such as thermoelectric power generation arrays and small energy recovery flywheel systems, to achieve efficient conversion and utilization of ≥99% of residual energy.

Future directions for discharge technology are shown in Figure 2.9.

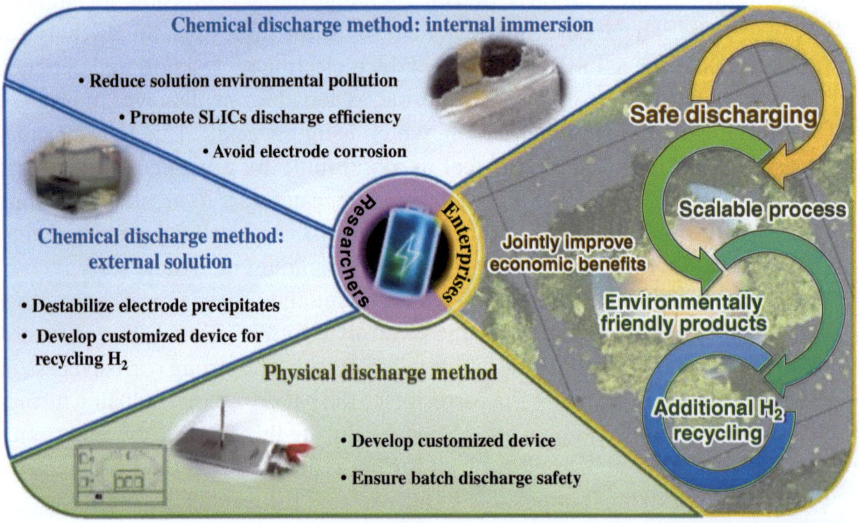

Figure 2.9 The prospects of future SLIC preprocessing discharge methods. *Source*: [47] / with permission of American Chemical Society.

2.4 Disassembly and Crushing

2.4.1 CMP Structure Battery Pack Disassembly

A battery pack consists of multiple single-cell batteries connected in series and parallel combinations, integrating subsystems such as thermal management, electrical control, and structural protection. It is a high-voltage energy storage device and serves as the core energy carrier for EVs and energy storage systems. When preprocessing spent batteries, the PACK should first be disassembled. The PACK generally comes in two types: CMP (Cell-Module-Pack) structure battery pack and CTP structure battery pack. The CMP structure battery pack is assembled into modules by cells, and then the modules are installed in the battery pack to form a three-level assembly mode of "cell-module-battery pack."[57] The CTP structure battery pack skips the standardized module stage, using adhesive to directly assemble the cells into the battery pack casing, forming a secondary assembly mode of "cell-battery pack." According to different structures, the disassembly method is divided into CMP structure battery pack disassembly and CTP structure battery pack disassembly.[57]

1. **CMP structure battery pack disassembly**
   The CMP structure battery pack is currently the most common type of battery pack available on the market. The disassembly method needs to be selected according to the type of casing. When disassembling battery packs that use bolted assembly connections, appropriate professional tools and equipment corresponding to the type and specifications of the bolts must be used. When disassembling battery packs that involve metal welding or encapsulated connections, professional cutting equipment must be used, and the cutting position and depth must be precisely controlled. When disassembling battery packs for embedded connections, professional mechanical cutting equipment must be used.[57]
2. **CTP structure battery pack disassembly**
   There is a lot of adhesives in the CTP battery pack, which can be divided into structural adhesive and thermal adhesive. Structural adhesives mainly include polyurethane structural adhesives, acrylic structural adhesives, silicone, epoxy structural adhesives, UV adhesives, and high-temperature-resistant hot melt adhesives. These adhesives are applied to the bottom of the battery cells and between the cells, tightly bonding and fixing different battery cells in parts such as the EV chassis. Thermally conductive adhesive mainly consists of a resin matrix (such as epoxy resin, silicone, polyurethane, etc.) and thermal conductive fillers (such as aluminum nitride, boron nitride, silicon nitride, alumina, magnesium oxide, zinc oxide, etc.). It is applied on the surface of the battery cells to achieve uniform heat dissipation. The bonding adhesive has high viscosity, strength, and elasticity, making it difficult to disassemble such battery packs directly. Therefore, before disassembly, the adhesive needs to be treated first. According to the different components and properties of the adhesive, the adhesive is often treated by low-temperature degradation and chemical degradation.

1. **Low-temperature degradation:** Generally, it involves processing under low temperatures (below −30 °C) for more than 48 hours, which weakens or eliminates the adhesive effect of the binder.
2. **Chemical degradation:** Generally achieved by soaking in chemical agents, which causes changes in the internal structure of the adhesive, thereby losing its stickiness.

The entire process of disassembling the battery pack must be carried out in a sealed device equipped with automatic temperature and smoke alarms. Once an accident occurs, immediately put it into the explosion-proof box to avoid safety risks. At present, PACK pack disassembly usually uses a crane to lift the battery pack onto the disassembly line, and the dust on the surface of the battery pack is cleaned by an industrial dust collector. Then use a disassembly tool to remove the bolts on the upper cover, the uppercase cover, the high-voltage wiring harness in the battery pack, the low-voltage wiring harness of each module, and the fixing bolts of each module, and take out the module. Finally, at the module disassembly workstation, remove the module cover, busbars, and module side panels, and take out the single-cell batteries.[6]

2.4.2 Battery Cell Collection

After battery pack disassembly, the cell will be obtained. According to different packaging methods, models, and materials of cells, there are three common battery collection methods: the packaging method collection, the model collection method, and the material collection method.

1. **Aggregation method of encapsulation**
   Currently, the common packaging methods for spent battery cells on the market are divided into hard case and soft pouch types. The hard cases come in two shapes: cylindrical and square, and they are encapsulated using steel or aluminum casings. Soft packaging is sealed using aluminum-plastic film and steel-plastic film. By determining the shape, different individual battery cells can be effectively categorized.
2. **Model aggregation method**
   To meet the requirements of different scenarios, there are also various models of battery cells. Common models of square batteries include 3578131 (indicating that the thickness of this square battery is approximately 3.5 mm, the width is about 78 mm, and the height is around 131 mm, same below), 3463110, 3845120, 366090, 3435165, 2453135, 385085, 344461, and 255480, among more than 20 types. Pouch cell (bag type) battery has a flexible shape, a variety of models, thickness is usually 3~10 mm, width is 30~200 mm, length is 50~300 mm. The common models of cylindrical batteries are 10440 (indicating that the cylindrical battery has a diameter of 10 mm and a height of 44 mm, the same below), 14500, 16340, 18650, 21700, 26650, 32560, 4680, etc.
3. **Material aggregation method**
   Currently, common spent batteries are mainly classified by their cathode materials into types such as LFP, ternary, and LCO. Ternary batteries, based on nickel

content, can be divided into $LiNi_{1/3}Co_{1/3}Mn_{1/3}O_2$ (NCM111), $LiNi_{0.5}Co_{0.2}Mn_{0.3}O_2$ (NCM523), $LiNi_{0.6}Co_{0.1}Mn_{0.3}O_2$ (NCM613), and $LiNi_{0.6}Co_{0.2}Mn_{0.2}O_2$ (NCM622). With the gradual retirement of high-endurance EVs, high-nickel ternary represented by $LiNi_{0.7}Co_{0.15}Mn_{0.15}O_2$ (NCM701515) and $LiNi_{0.8}Co_{0.1}Mn_{0.1}O_2$ (NCM811) will gradually enter the recycling market. By collecting different materials, single-cell batteries of different materials can be effectively collected, which provides convenience for subsequent efficient crushing and sorting.[6]

In the early days, the collection of single-cell batteries was mainly carried out by manual sorting. With the rise and development of information technology and intelligent identification technology, the spent battery sorting line can intelligently identify and machine sort the size and shape of cells, so as to realize packaging mode and model collection.[54] At the same time, with the formal implementation of "Interim Provisions on the Traceability Management of New Energy Vehicle Power Battery Recycling" in 2018, the surface of cell shall be gradually marked with a specific 24-bit battery code, and the manufacturer, product type, battery type, specification, traceability information, production date, sequence, and other information of cell shall be identified in detail. Through the identification of the battery code, the spent battery cells can be sorted and collected efficiently and quickly. For example, the fifth digit of the code is the battery type code, which represents the battery type. For a multicomponent mixed material system, the code is written with the material component with the largest content, and there are two or more material components with the same content, and the code is written with the material component with poorer safety. Codes A, B, C, D, E, F, G, and Z represent nickel–hydrogen battery, LFP battery, LMO battery, LCO battery, ternary material battery, super capacitor, LTO battery, etc. respectively. Effective identification of the cell code allows material collection.

2.4.3 Fragmentation

Crushing typically involves using a series of crushing equipment to process batteries, aiming to separate the different components within the battery. Depending on the method used, it can be categorized into dry crushing and wet crushing.

Wet crushing, unlike dry crushing, needs to be carried out in water or other solutions. Since wet crushing operates within a solution system, it can effectively avoid issues of dry powder splashing, suppress dust generation, and help eliminate dust emissions as well as equipment blockages during dry crushing. At the same time, the solution used in crushing can also be used as a detergent and flame retardant to absorb the heat generated in the crushing process. In addition, in this process, lithium is converted to an inert state and no longer reacts with the electrolyte, reducing the risk of fire and explosion when the battery is broken. By using wet crushing to process spent batteries, the working noise during the crushing process is low, and energy consumption is low, which helps to shorten the processing cycle and enhance operational safety.[58] However, wet crushing allows adhesives, electrolytes, and other toxic substances from spent batteries to be released from the decomposed batteries, forming highly toxic and highly polluting wastewater, which will further increase the operating costs of wastewater treatment. In addition, using wet crushing, copper and aluminum foils have excessive crushing phenomenon, which makes the impurities in the polar powder recovered by subsequent sorting usually higher than dry crushing, which is not conducive to subsequent direct repair or hydrometallurgy.[59]

Dry crushing, generally involves processing spent batteries in an air or inert atmosphere. During the preprocessing stage, spent batteries generally need to undergo multistage crushing. For discharged batteries, they can be torn apart in the first stage to break the battery casing, allowing the components to initially separate. The iron shells and separators obtained after tearing can be recycled through magnetic separation or air flow sorting methods.[45] Subsequently, the cathode sheet and anode sheet, along with other battery components, can be further crushed or ground to specific sizes, allowing the active material powder to separate from the copper and aluminum foils, facilitating the subsequent separation of valuable components. However, during the dry crushing process, fragments of the battery's cathodes and anodes can form micro-short circuits, leading to a sharp increase in heat release. Friction caused by breakage leads to temperature rise, making the organic substances in the battery prone to decomposition at high temperatures, and even combustion and explosion. During the fragmentation period, the fragmentation temperature is closely related to the battery charge level. The higher the battery charge level, the higher the fragmentation temperature, and the greater the risk of fire or explosion. Therefore, deeply discharging batteries before they break is an important prerequisite for safe production.[60] Moreover, when performing dry crushing, it is necessary to collect and treat the electrolyte and dust, and an integrated dust control system is required to prevent dust pollution or explosions. After discharging, the battery's capacity for dry crushing treatment is large, making it convenient for industrial production and successfully achieving industrial application.[61]

1. **Charged fragmentation**
   Charged crushing, that is, under the protection of inert gas such as nitrogen or low-temperature environment, the charged spent battery is directly crushed without discharging. In order to enhance the safety of operation, nitrogen is usually continuously introduced to maintain the oxygen-poor environment during charged fragmentation, or the activation energy conditions in the three elements of combustion are blocked by low-temperature thermodynamic effect to reduce the possibility of fire and explosion during charged battery fragmentation.[24] At the same time, crushing under inert gases such as Ar, N_2, and CO_2 can significantly reduce the activity of lithium and its by-products and prevent spontaneous combustion or the formation of toxic gases. This crushing method not only avoids the cumbersome steps of traditional discharge treatment, greatly improves production efficiency, but also reduces potential safety hazards from the source, providing a strong guarantee for the smooth progress of the entire recovery process. In this process, the spent battery first enters the shredder, and under the protection of nitrogen, the spent battery is

successfully crushed into small particles with a particle size of ≤30 mm through the precise shearing of the two-stage shredder head. After that, the components in the battery are completely dissociated through multistage crushing process. However, a large amount of inert gas is consumed during charged crushing, which increases the overall cost. At the same time, if the oxygen content or temperature in the crushing system is too high, there is still a risk that the cathode and anode materials will catch fire during the crushing process.[44]

2. **Freezing crushing**
   Cryogenic crushing: liquid nitrogen is used to apply extremely low temperatures (−175 ~ −200 °C) to spent batteries to freeze the electrolyte and reduce the conductivity and electrode material activity of lithium-ion batteries, thus avoiding thermal runaway, combustion, and explosion risks during the crushing process. At the same time, too low temperature makes the battery, electrode, and other components brittle, which is conducive to the subsequent crushing process. After applying low temperatures, the aforementioned dry crushing method can be used to crush spent batteries. Crushing can effectively avoid safety risks caused by highly active lithium during battery disassembly. This technology can handle a large number of spent batteries at the same time, especially suitable for handling high-capacity batteries.[61] However, due to the need to use a large amount of coolant such as liquid nitrogen in the crushing process, the requirements for capital and equipment are very high. The risk of fire and explosion in battery crushing is high, and the risk of fire and explosion in battery crushing is high.

First of all, the lithium element contained in the spent battery is highly active. In the process of crushing, the cathode and anode fragments of the battery are connected, forming a micro-short circuit, resulting in a sharp increase in heat release. The higher the battery charge, the higher the amount of heat released when a micro-short is formed. At the same time, different components such as cathode, anode, and steel sheet in the crushing process will also induce high heat generation due to mechanical friction, causing explosion. Although working in an inert atmosphere can, to some extent, avoid the flammable and explosive effects of oxygen on the components of spent battery materials. However, in actual production, it is impossible to completely create an absolutely oxygen-free environment, which will also lead to a high risk of battery fires and explosions.[60]

Secondly, the battery contains more combustibles, such as conductive agents, binders, graphite, separators, etc., and the high mechanical heat generated during the crushing process is easy to cause combustibles to catch fire. There is also a residual volatile electrolyte in the spent battery. The electrolyte contains carbonate solvents with low flash points. In severe crushing treatment, it is easy to cause fire and explosion due to friction heating.

In addition, in the crushing process of the battery equipped with ternary cathode material, because the ternary cathode contains a large amount of oxygen element and has quite strong activity, strong crushing treatment will induce the ternary cathode to spontaneously release oxygen, increasing the risk of fire and explosion of the battery.

Battery disassembly and crushing technology development needs: the development of waste lithium-ion battery disassembly and crushing technology needs to be promoted around the core goals of safety, efficiency, green, and intelligence.[24]

In terms of safety, it is necessary to overcome the bottleneck of electric discharge fragmentation risks by combining a nitrogen inert environment with multiple sensor protections to achieve an accident rate of less than 0.1%. Additionally, integrate low-temperature pyrolysis for electrolyte recovery to avoid the generation of highly toxic HF.

In terms of efficiency, developing flexible multilevel crushing systems and artificial intelligence (AI) sorting technology that is compatible with cylindrical, square, pouch, and other heterogeneous batteries ensures that the purity of copper and aluminum recovery is ≥98%, the recovery rate of black powder is ≥98%, and the magnetic impurities are <50 ppm, meeting the requirements for direct material regeneration.

In terms of environmental protection, a comprehensive pollution control system must be established: RTO incineration combined with alkali scrubbing towers achieves VOCs removal rate of ≥95%, closed-loop water circulation ensures zero wastewater discharge, and thermal cracking residues solidify heavy metals >99.99% for use in construction materials.[58]

At the policy coordination level, relying on battery full lifecycle traceability (RFID/barcode) to match the optimal process, and promoting the construction of a network between car manufacturers and dismantling companies through regional policies (such as Chongqing's goal of 90% recycling coverage by 2027), while using value-added tax incentives (refund 50% immediately) to reduce operational costs.

Future technology needs to focus on upgrading low-carbon processes (reducing thermal decomposition residual heat consumption by 80%) and modular production lines (cutting investment costs by 40%), driving the recycling industry to shift from a cost center to a resource profit center.

A comparison of the various fragmentation methods is provided in Table 2.2.

Table 2.2 Comparison of various fragmentation methods.

Technology type	Advantage	Main risks	Applicable scenarios
Dry crushing	High processing capacity Industrial maturity The purity of the copper powder is relatively high	Dust explosion risk Micro-short circuit overheating	Low-power battery (SOC < 30%)
Wet crushing	Dust-free ($PM_{2.5}$ < 10 ug/m^2) Flame-retardant cooling	The wastewater has high toxicity Metal foil over-crushed	Small soft-pack battery
Electrostatic fragmentation	No discharge (saving 6–8 hours) Inert gas protection (O_2 < 1%)	The idling consumption is high The oxygen control system requirements are high.	High-value batteries (NCM/NCA)

Technology type	Advantage	Main risks	Applicable scenarios
Freeze-fracture	Zero-burn (−196 °C) Material activity inhibition (Li passivation)	The cost of liquid ammonia is high The equipment must withstand low temperatures	High-capacity battery (≥100 kWh batch)

2.5 Material Sorting

After fragmentation and extreme powder separation, aluminum foil powder, anode copper foil powder, anode powder, cathode powder, binder, conductive agent, metal shell powder, separator, and other components still exist in a mixed form. To separate and enrich each component, different sorting methods need to be used to process the battery components.[62] Based on the differences in physical properties such as particle size, shape, density, electrical conductivity, magnetic permeability, and wettability of each component, the sorting methods that can be used include screening, pneumatic separation, electrostatic separation, eddy current separation, magnetic separation, and flotation.[63]

2.5.1 Screening and Pneumatic Separation

Screening, which separates objects of different particle sizes using sieves with different mesh sizes. It is a traditional technique used in the recycling and utilization of solid waste. The ease of sieving materials of different particle sizes varies; the smaller the material particles, the easier they pass through the sieve holes. When the material and sieve hole sizes are similar, sieving becomes more difficult. During the recycling process of used batteries, screening usually takes place after the crushing stage to separate materials of different particle sizes resulting from the crushing. The sieving separation process is schematically shown in Figure 2.10. When the battery is broken, different components of the battery have good selective breaking, that is, different components often exhibit different particle size characteristics, such as most copper components are concentrated in the +2 mm particle size range, LCO, and graphite are concentrated in the −0.25 mm range, and a small amount of aluminum and copper are concentrated in the 2 ~ +0.25 mm range (Figure 2.11). By screening, fine powder and metal components can be effectively separated. By adopting a process of crushing, screening, ultrasonic/water bath treatment, and rescreening for spent batteries, the separator can be separated by screening after the initial crushing. After ultrasonic washing and water bath treatment, the active material powder can be stripped from the current collector, and then screening is used to separate the active material powder from aluminum and copper foils, achieving an active material separation efficiency of 92%.[54]

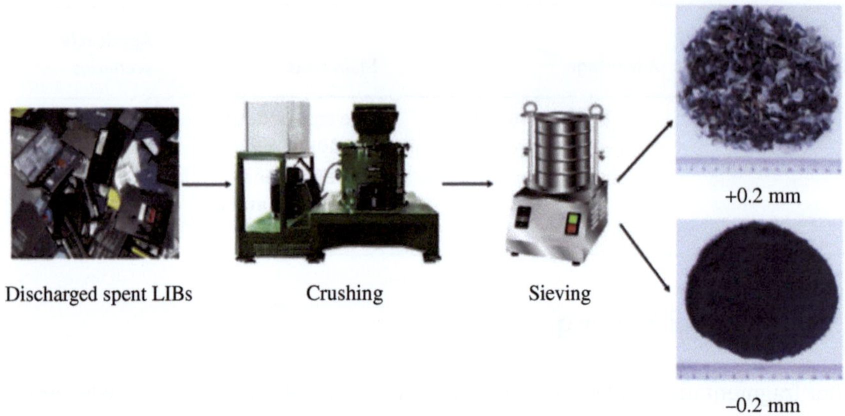

Figure 2.10 Screen the broken and released material to separate the coarse product (+0.2 mm) including metal shells, organic films, copper/aluminum foils, and other plastics from the fine product (−0.2 mm) including graphite and $LiCoO_2$ particles. *Source*: [64] / with permission of Elsevier.

Figure 2.11 In particle size separation, the eight particle size fractions of the waste LIB after crushing. *Source*: [6] / with permission of Springer Nature.

Pneumatic separation, also known as air separation, uses an air classifier to separate different components based on particle size, shape, and density. A schematic of the pneumatic separation process is shown in Figure 2.12. During air flow sorting, air is used as the sorting medium. Under the effect of the airflow, light materials are carried away from the top or transported to a greater distance horizontally, while heavy materials settle due to the inability of the airflow to support them, thus achieving separation between different materials. Generally speaking, as the airflow speed increases, the yield of light products increases, while the yield of heavy products decreases. Based on this, in the broken materials of spent batteries, the density of plastic separators is small, and they are easily sorted under the action of air flow. At the same time, aluminum powder is lighter than copper powder, and the separation of the two can be achieved by controlling the air flow size and

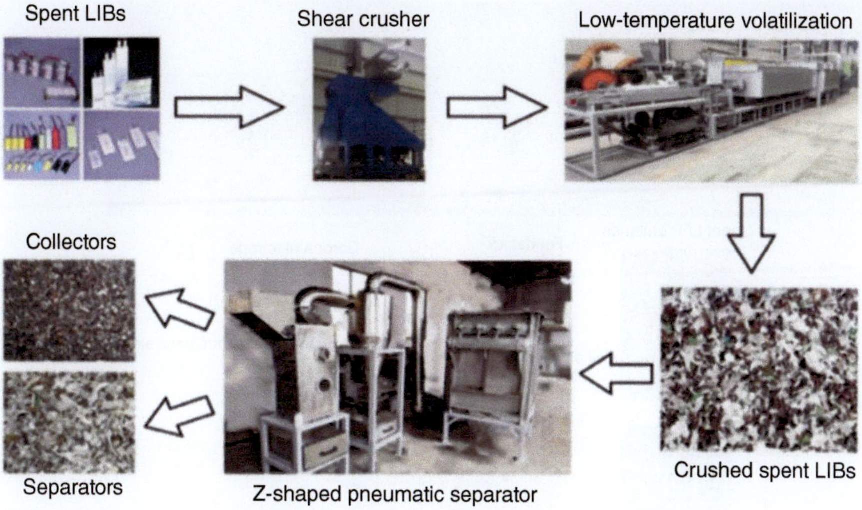

Figure 2.12 Diagram of pneumatic separation of waste LIBs in industrial processes. *Source*: [6] / with permission of Springer Nature.

feeding speed. When the size of crushed material is 3 ~ 4 cm and the velocity of air flow is 6.96 ~ 7.8 m/s, the separation efficiency between separator and collector can reach almost 100%. Pneumatic separation efficiency and accuracy are high, and the operation is simple. At present, it has been widely used in the recovery pretreatment process of spent batteries. Its shortcomings are that the air separation effect on irregular shapes, similar density of materials is not good, and the air separator has high energy consumption, complex maintenance, and needs regular cleaning and replacement of parts.

2.5.2 Electromagnetic Separation

As shown in Figure 2.13, electrostatic separation, based on the difference of conductivity, friction characteristics, dielectric constant, and other properties among spent battery components, makes the particles with different characteristics have different movement trajectories under the action of different forces through an applied electric field, so as to realize the separation between different materials. In lithium-ion batteries, in addition to metals, plastic polymers and other nonmetals also account for a certain proportion, the difference in conductivity between metals and nonmetals is large, electrostatic separation can be used for the separation between metal parts and plastic polymers. After applying high-voltage DC electricity, the corona electrode ionizes the surrounding air, causing negatively charged ions to fly toward the transfer roller electrode, forming an ionized area between the corona electrode and the transfer roller electrode. When the feed enters the ionization area, metal conductor particles, and nonconductor particles such as plastic polymers enter the charging process. The negative charge acquired by conductive particles is carried away by the roller electrode, leaving only positive charges on the surface. Nonconductive particles such as plastics, which have strong dielectric

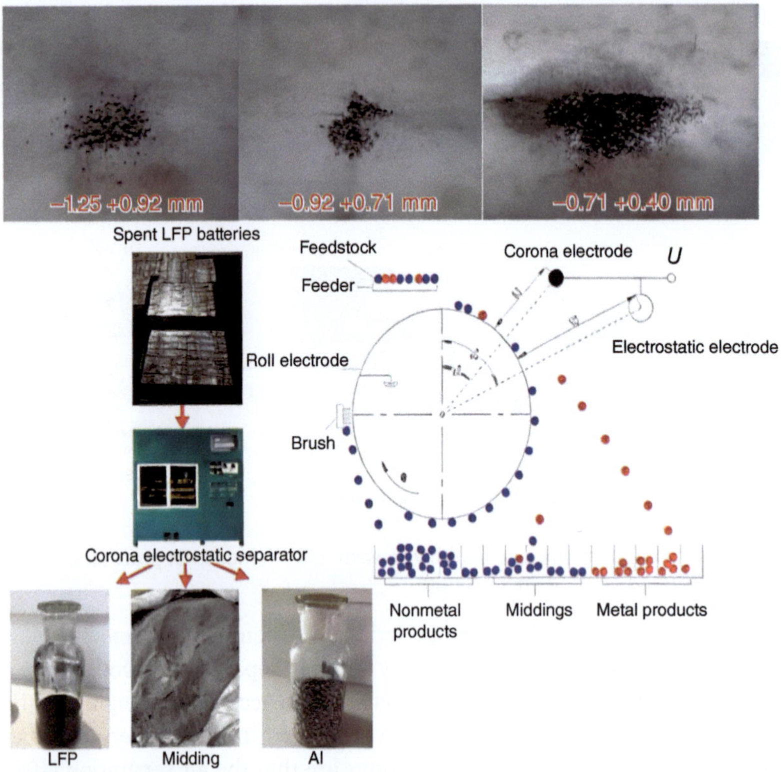

Figure 2.13 Particle mixtures and electrostatic separators (ES) in electrostatic separation samples. *Source*: [6] / with permission of Springer Nature.

properties, find it difficult for their negative charges to be carried away, resulting in a negatively charged surface. Therefore, metal particles and nonmetallic particles exhibit different motion states in the electrostatic field generated by the electrostatic electrodes.

Conductor particles with positive charge are repelled and separated by the rotating roller electrode under the compound action of electric field force, centrifugal force, and self-gravity, and finally fall into the conductor particle collection area due to oblique throwing motion in the electrostatic field. Nonconductive particles, due to their own negative charge, adhere to the surface of the rotating roller electrode. As the roller electrode rotates, they eventually fall into the nonconductive particle collection area. Particles between conductors and nonconductors fall into the intermediate product collection area. Electrostatic separation has high efficiency, but when the moisture content of the material is high, its ability to become charged decreases, resulting in poorer separation performance.[65]

The principle of eddy current separation technology is shown in Figure 2.14. It based on the difference of conductivity between the broken components of spent batteries, Lorentz force is used to change the path of different materials to achieve the separation of components. In this process, the conductor in the high-frequency alternating magnetic field can produce induced current, and the current can produce a magnetic

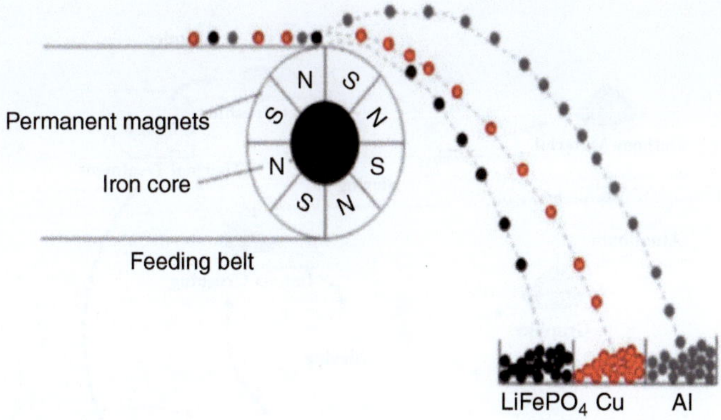

Figure 2.14 Eddy current separation technology. *Source*: [6] / with permission of Springer Nature.

field, thereby generating the Lorentz force. Eddy current sorter will produce high-frequency alternating magnetic field on the surface of the sorter roller when it works. When the metal particles move into the strong magnetic field separation area, an induced alternating current is generated inside the metal particles, and the alternating current excites a new magnetic field opposite to the original magnetic field around the particles, generating mutual repulsion force, and the metal particles are thrown forward under the repulsion force, thereby realizing the separation of metal and nonmetal objects. In the process of sorting valuable components of failed lithium-ion batteries, due to the different magnetic field induction degrees of copper, aluminum, and waste electrode powder, the electrode powder, copper and aluminum foils will show obviously different motion paths by controlling the particle size of the broken product and adjusting the rotation speed of the magnetic roller, so as to separate the electrode powder, aluminum and copper foils from each other in the spent lithium-ion batteries.[66]

Magnetic separation, using the different permeability between different components, by controlling the magnetic field intensity, so that the components with high permeability are selectively attracted by the magnetic medium, so as to achieve the purpose of separating materials. The magnetic separation process flow is illustrated in Figure 2.15. In this process, the material enters the nonuniform magnetic field of the magnetic separator, and at the same time, it is affected by magnetic force and competitive force, and the resultant force determines the movement track of the particles. The specific magnetic force of the material in the magnetic field depends on the magnetism of the material itself, the magnetic field strength, and the magnetic field gradient. When the magnetic field is constant, the higher the specific susceptibility of the object, the greater the specific magnetic force, the material receives in the magnetic field, and the easier it is to be attracted to the magnetic medium. Among the components of spent lithium-ion batteries, magnetic parts mainly come from iron-based materials, especially the steel casing of cylindrical batteries.[44] Additionally, LFP, LMO, and NCM cathode are paramagnetic solids, while LCO, graphite, and copper and aluminum foils can be considered diamagnetic solids. In theory, magnetic separation can be used to isolate cathode materials (excluding LCO), graphite, and copper and

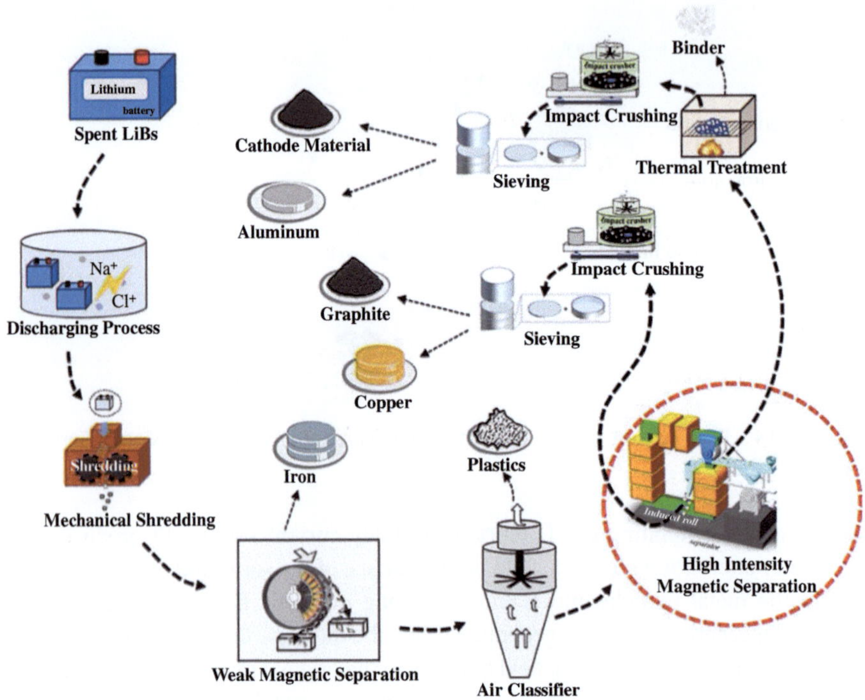

Figure 2.15 Flowchart of the process for recycling used lithium batteries using magnetic separation. *Source*: [67] / with permission of Elsevier.

aluminum foils from spent lithium-ion battery components. However, magnetic separation requires a high degree of particle size for the separated substances. When using magnetic separation to separate cathode and anode materials, due to phenomena such as mechanical entrapment and magnetic attraction entrapment, it is still difficult to achieve complete separation of the cathode and anode materials. Therefore, magnetic separation is currently mainly used to separate steel casings from battery components.

2.5.3 Flotation

Flotation, which achieves separation of components in end-of-life batteries by utilizing differences in wettability. The wettability of raw materials refers to the ability of a liquid to maintain contact with a solid surface, which is the most critical factor in the flotation process. Significant differences in the wettability of components contribute to enhancing the efficiency of flotation separation.[68] A schematic of the flotation separation process is shown in Figure 2.16. In the process of handling crushed materials from spent batteries, due to the significant difference in wettability between the cathode material and the graphite anode material, theoretically, flotation can be used to separate the cathode material from the anode material. That is, graphite has strong hydrophobicity and adheres to bubbles, while the more hydrophilic cathode material settles at the bottom.[45] However, for spent battery components, in order to improve the conductivity of the material, a carbon coating layer is

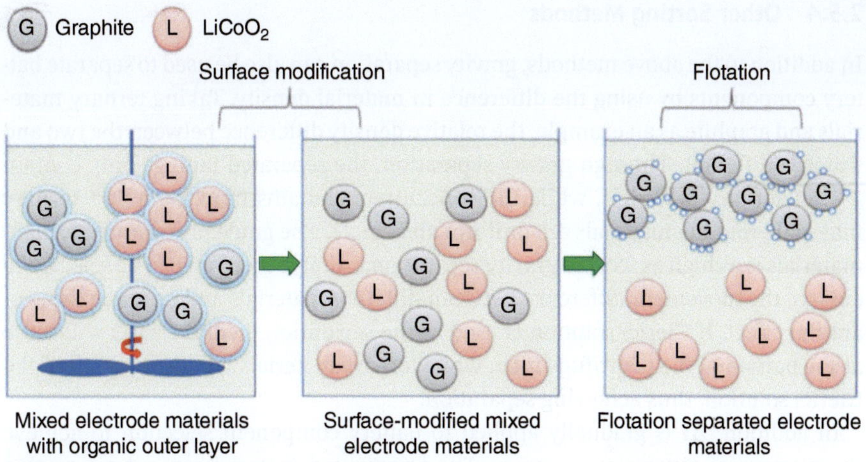

Figure 2.16 A schematic diagram illustrating the use of Fenton's reagent-assisted flotation method to separate graphite and LCO.[54]

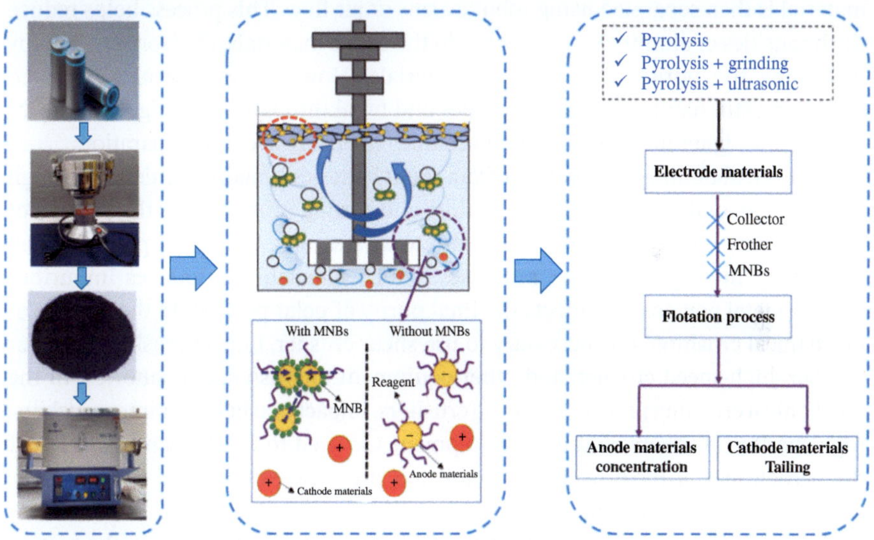

Figure 2.17 A schematic diagram of pyrolysis pretreatment combined with MNBs auxiliary flotation recovery method. *Source*: [69] / with permission of Elsevier.

usually indicated on the cathode material, so the cathode material also exhibits wettability similar to graphite, making the flotation effect poor. To improve the flotation separation effect, processes such as Fenton oxidation enhancement, mechanical grinding enhancement, and thermal treatment enhancement can remove binders, expand the wettability difference between cathode materials and anode materials. The pyrolysis-pretreatment combined with flotation recovery method is illustrated in Figure 2.17. However, the combined method also will increase the process flow and affect production efficiency. Therefore, flotation is rarely used in practice.[66]

2.5.4 Other Sorting Methods

In addition to the above methods, gravity separation can also be used to separate battery components by using the difference in material density. Taking ternary materials and graphite as an example, the relative density difference between the two and water is different. Through gravity separation, the separated tailings only contain 3 ~ 5% ternary materials, while the concentrate contains more than 50% ternary materials, and the materials are uniform and single. The grade of separated ternary materials is as high as 98%. In gravity separation, suitable medium can be selected to enlarge the movement difference between different materials and enhance the separation effect. If Clerici solution is used as the separation medium, LCO will settle at the bottom of the centrifuge tube, while ternary materials will float on top of the Clerici solution, thus achieving separation.

In addition, AI is gradually applied to battery component selection to achieve the purpose of separation accuracy. The fragmented material can enter a sorting system assisted by computer vision and machine learning, such as using multispectral imaging to identify metals like aluminum and copper and active particles, or near-infrared and X-ray detection for rapid identification of material properties. The material is then separated using robotic arms or air flow. This process helps reduce the impurities of aluminum and copper in the active materials after sorting, thereby improving the purity of the active materials. However, intelligent automated processing still requires further upgrades and breakthroughs in AI algorithm technology and higher image recognition accuracy to ensure accurate separation.

However, obtaining polar powder from spent batteries is mainly achieved through battery disassembly, mechanical crushing, and sorting processes. In the process of PACK battery disassembly, it is necessary to disassemble the PACK package shell by cutting, milling, and other processes, which inevitably introduces impurities such as metal powder and affects the final purity of polar powder. In the process of mechanical crushing, it is necessary to use shear crusher, impact crusher, hammer crusher, high-speed crusher, and other equipment to crush spent batteries. In the mechanical crushing process, shearing crushers, impact crushers, hammer crushers, high-speed pulverizers, and other equipment are used to crush spent batteries. If the particle size of the broken material is too large, the cathode and anode sheets, steel sheets, separators, and other components may be mixed together, ultimately leading to the introduction of impurities from non-active substances in the battery into the recycling process, affecting both the recycling efficiency and the quality of the final product. However, if the particle size is too small, a large amount of dust will be formed, which makes the difference of separated particles in the subsequent separation process not obvious and the separation incomplete. During the sorting process, the screening process has poor separation effect on materials of similar size, and the pneumatic sorting process has poor separation effect on irregularly shaped materials with similar density. Electrostatic separation is not effective for materials with similar conductive properties, dielectric performance, and moisture content; magnetic separation is not effective for materials with similar magnetic properties, mechanical impurities, and attracted impurities; and flotation has poor separation effects on materials with similar wettability properties. This results in the inability to precisely separate the cathode and anode powders in the final product, leading to a mixture of cathode and anode materials containing impurities such as aluminum and copper, which is unfavorable for the classification, recycling, and utilization of cathode and anode materials.[69]

Battery sorting technology demand analysis: The future development of spent lithium-ion battery sorting technologies must focus on four core directions—safety enhancement, intelligent upgrading, environmental sustainability, and policy coordination:

In terms of safety, it is urgently necessary to popularize inert environment protection systems (nitrogen protection, oxygen content $\leq 2\%$), combined with multilevel monitoring of temperature, flame, and oxygen concentration, as well as low-temperature pyrolysis technology (100–200 °C) to achieve an accident rate of less than 0.1% during electric breaking and efficiently recover electrolytes to avoid HF pollution.

In terms of smart upgrades, multimodal AI recognition (visual + spectral analysis) is used to enhance sorting accuracy, achieving a purity of copper and aluminum foil $\geq 98\%$ and a black powder recovery rate $\geq 99\%$; modular flexible production lines are developed to accommodate cylindrical, square, pouch, and other heterogeneous batteries, reducing process switching time to 10 minutes.

In terms of deepening environmental protection, a full-process pollution control system is being constructed: RTO incineration combined with alkali scrubbers ensures VOC removal efficiency of $\geq 95\%$; closed-loop water circulation achieves zero wastewater discharge; promoting waste heat recovery (reducing consumption by 80%) and room temperature hydrometallurgy (such as sulfuric acid leaching reducing consumption by 60%) and other low-carbon processes.

At the policy coordination level, leveraging battery full lifecycle traceability (using RFID/barcode labeling for components), automatically matching sorting processes (such as ternary 60–80 °C/LFP 20–30 °C water immersion), through regional collaboration networks (such as achieving 90% county coverage in Chongqing by 2027), integrating resources between automakers and recycling companies, and adding VAT incentives (refund of 50% on-the-spot) to reduce operational costs.

Future technology will evolve toward the goal of "zero safety risk, zero sorting error, and zero pollution emission," driving the industry from cost center to resource profit center.

2.6 Summary

Based on current technological limitations and market demand, future green high-value recycling technologies for spent lithium-ion batteries will focus on the following core areas, gradually achieving comprehensive improvements in resource recycling efficiency, environmental friendliness, and economic value.

In terms of detection: The primary objective is to achieve intelligent detection by leveraging machine learning algorithms to establish a smart battery inspection system. This system integrates multiple detection methodologies and data analytics technologies, enabling real-time data acquisition, processing, and analysis. Using AI algorithms, the health status of batteries and potential risks can be automatically identified, improving detection efficiency and accuracy. Secondly, achieve universal detection: the developed detection technology should have extremely strong universality, capable of providing uniform evaluation for different types of batteries. By optimizing the detection model through iterative techniques, ensure long-term operational stability. Thirdly, to achieve efficient detection: develop high-precision, high-sensitivity detection equipment that can quickly and accurately evaluate battery safety, SOH, and life. Moreover, detection technology should enable nondestructive testing of the internal state of batteries. Through mechanism research, deeply analyze the aging mechanisms of batteries, explore easily characterized feature parameters that are highly correlated with battery status, and enhance the detection of material realization mechanisms in batteries. In addition, the detection time needs to be reduced from several minutes to just a few seconds to meet the demands of large-scale industrial testing. Finally, achieve standardized testing: establish standardized testing procedures and norms to ensure the reliability and consistency of test results. The standardized testing process should include testing methods, testing parameters, data processing, and result determination, providing a unified technical basis for the second-life use.

In terms of precise disassembly: Intelligent and precise disassembly of PACK packs, achieving automatic recognition of spent PACK battery packs and rapid adaptation to different PACK structures. Realize the battery disassembly process "AI + multi-structure compatibility + full automatic operation," all mainstream PACK structures can be fully automatic, efficient, refined disassembly by AI-driven robots, and achieve PACK package automatic identification and classification accuracy rate of 100%, full-process intelligent disassembly coverage rate of 100%, average disassembly time (standard PACK) $\leq$ 1 minute/piece.

In terms of discharge: Deep discharge of cell, residual energy utilization, establishment of green, efficient, and safe automatic discharge system, and realization of "zero pollution and zero emission" in the discharge process. At the same time, develop and integrate multipath residual energy utilization platform to maximize residual energy utilization rate, and achieve discharge depth of 100%, residual energy utilization rate $\geq$ 99%.

In terms of crushing and sorting: Battery safe crushing, efficient sorting, combined with AI technology, machine vision technology, adaptive robot arm linkage control technology, and other technologies, to achieve intelligent, fine and safe crushing and efficient sorting, reduce manual processing, and obtain high-purity battery powder, and achieve the recovery rate of polar powder/copper/aluminum foils more than 99.5%, the recovered polar powder impurity content is less than 50 ppm.

References

1 Y. Zhou, H. Wang and J. Wang, *Batteries & Supercaps*, 2024, **7, 00028.**

2 F. Zhu, X. Du, J. Lei, L. Audisio and D. Sypeck, *International Journal of Crashworthiness*, 2020, **26**, 598–607.

3 L. Chen, T. Gu, J. Mi, Y. Li, K. Yang, J. Ma, X. An, Y. Jiang, D. Zhang, X. Cheng, S. Guo, Z. Han, T. Hou, Y. Cao, M. Liu, W. Lv, Y.-B. He and F. Kang, *Nature Communications*, 2025, **16, 58689.**

4 Y. Shi, M. Wang, Z. Yang, L. Meng, X. Huang, Y. Wang, Z. Li, Y. Lu, Y. Cai and J. Wang, *Applied Surface Science*, 2025, **714, 164449.**

5 M. Hwang, J. H. Choi, S. Lee, J. Hwang, S. Park, S. Choi, M. Kim, H. Lim, H. Lim, M. Oh, S. Song, G. Shin, M. Park, Y. Kim, D.-H. Seo and P. Oh, *Joule*, 2025, DOI: 10.1016/j.joule.2025.102103.

6 S.-X. Yan, Y.-Z. Jiang, X.-P. Chen, L. Yuan, T.-T. Min, Y. Cao, W.-L. Peng and T. Zhou, *Rare Metals*, 2023, **43**, 915–941.

7 S. Hong, D. Ma, W. Zeng, J. Shi, Y. Liu, L. Yang, Y. Fan, Y. Liu, C. Yang and B. Hong, *Journal of Power Sources*, 2025, **629, 236013.**

8 G. Zheng, L. Tan, G. Tian, B. Yang, Q. Li and G. Sun, *Thin-Walled Structures*, 2023, **192, 110902.**

9 D. Qin, P. Wang, T. Wang and J. Chen, *Sustainability*, 2023, **15, 108414.**

10 B. Wang, Y. Dong, Z. Zeng, H. Lei, Y. Yang and P. Ge, *Journal of Power Sources*, 2024, **621, 235299.**

11 Z. Fang, Q. Duan, Q. Peng, Z. Wei, H. Cao, J. Sun and Q. Wang, *Journal of Cleaner Production*, 2022, **359, 132116.**

12 Z. Zhang, X. Li, L. Gong, Y. Li, J. Qin and P. Tan, *Energy & Fuels*, 2023, **37**, 9713–9721.

13 J. Jiang, X. Liu and R. Yang, *Energy Storage Materials*, 2025, 82, 104560.

14 Y. Li, P. Zhao and B. Shen, *Science of the Total Environment*, 2024, **951, 175459.**

15 L. Sheng, C. Zhang, J. Xu, X. Zhang, X. Wang and Z. Zhang, *Applied Thermal Engineering*, 2024, **256, 124081.**

16 J. Zhong, S. Feng, R. Mao, X. Wang, Z. Lei, R. Bai and C. Yan, *Journal of Power Sources*, 2025, **650, 237484.**

17 P. R. Tete, M. M. Gupta and S. S. Joshi, *Journal of Energy Storage*, 2022, **48, 104041.**

18 E. Lübke, L. Helfen, P. Cook, M. Mirolo, V. Vinci, O. Korjus, B. Fuchsbichler, S. Koller, R. Brunner, J. Drnec and S. Lyonnard, *Energy & Environmental Science*, 2024, **17**, 5048–5059.

19 M. Ahmadian-Elmi and P. Zhao, *Batteries*, 2024, **10, 020050.**

20 Y. Song, M. Bulla, H. Patanwala and E. Sahraei, *Journal of Energy Storage*, 2024, **94, 112371.**

21 J.-H. Seok, S. Lee, D.-A. Lim, K. H. Ahn, C. H. Lee, K. Kim and D.-W. Kim, *Journal of Materials Chemistry A*, 2025, **13**, 13976–13987.

22 Y. Cao, J. Li, H. Ji, X. Wei, G. Zhou and H.-M. Cheng, *Energy Storage Materials*, 2024, **70, 103475.**

23 X. Li, S. Liu, J. Yang, Z. He, J. Zheng and Y. Li, *Energy Storage Materials*, 2023, **55**, 606–630.

24 S. Wu, C. Hong, N. Kaden, P. Gümbel, D. M. Nguyen, C. Wacker and K. Dröder, *Journal of Energy Storage*, 2025, **131, 115723.**

25 L. Wildfeuer, N. Wassiliadis, A. Karger, F. Bauer and M. Lienkamp, *Journal of Energy Storage*, 2022, **48, 103909.**

26 M. Mousavi and A. H. Gandomi, *Construction and Building Materials*, 2021, **307, 124999**.

27 L. Giammichele, V. D'Alessandro, M. Falone and R. Ricci, *Applied Thermal Engineering*, 2022, **205, 117974**.

28 S. Ni, S. Lama, Y.-J. Lee and J.-H. Kim, *Arabian Journal for Science and Engineering*, 2024, **50**, 2527–2540.

29 K. Sato, A. Tamai, K. Ohara, H. Kiuchi and E. Matsubara, *Journal of Power Sources*, 2022, **535, 213399**.

30 Y. Shen, J. Zhang, S. Chen, K. Qu, Z. Yang and Y. Peng, *Nano Energy*, 2025, **140, 111065**.

31 O. Klvač, D. Trochta, L. Novák, P. Priecel, M. Bornhöfft, T. Kazda and Z. Liu, *Energy Storage Materials*, 2025, **81, 104516**.

32 Y. He, K. Jia, Z. Piao, Z. Cao, M. Zhang, P. Li, Z. Li, Z. Jiang, G. Yang, H. Xi, G. Zhou, W. Tang, Z. Qu, R. V. Kumar, S. Ding and K. Xi, *Angewandte Chemie International Edition*, 2025, **64, 22610**.

33 X. Lai, P. Ke, Y. Zheng, J. Zhu, E. Cheng, B. Tang, K. Shen and T. Sun, *Journal of Energy Storage*, 2024, **104, 114563**.

34 M. M. Camboim, A. C. Moreira, M. d. F. N. C. Rosolem, R. F. Beck, V. T. Arioli, C. Omae and H. Ding, *Journal of Energy Storage*, 2024, **78, 110063**.

35 D. Fang, F. Wu, X. Zhang, X. Zhang, Y. Ye, R. Chen and L. Li, *Angewandte Chemie International Edition*, 2025, **64, 02176**.

36 Q. Li, R. Song and Y. Wei, *Journal of Energy Storage*, 2025, **118, 116078**.

37 C. Armenta-Deu, *Frontiers in Energy Research*, 2025, **13, 1573972**.

38 X. Wan, X. Xu, F. Li, X. Song, C. Peng and J. Liu, *Small Structures*, 2023, **5, 2300196**.

39 G. Qian, X. Chen, H. Lin and L. Yang, *Cell Reports Physical Science*, 2024, **5, 102153**.

40 M. J. Alenjareghi, S. Keivanpour, Y. A. Chinniah and S. Jocelyn, *Journal of Intelligent Manufacturing*, 2024, 36, 5563–5591.

41 K. Bhaskar, A. Kumar, J. Bunce, J. Pressman, N. Burkell, N. Miller and C. D. Rahn, *Applied Energy*, 2025, **380, 125087**.

42 Z. Chen, R. Feng, W. Wang, S. Tu, Y. Hu, X. Wang, R. Zhan, J. Wang, J. Zhao, S. Liu, L. Fu and Y. Sun, *Nature Communications*, 2023, **14, 4648**.

43 M. E. Carbonó dela Rosa, G. Velasco Herrera, R. Nava, E. Quiroga González, R. Sosa Echeverría, P. Sánchez Álvarez, J. Gandarilla Ibarra and V. M. Velasco Herrera, *Energies*, 2023, **16, 1073**.

44 U. Kar, A. Fahimi, T. Suponik, M. Kaya and P. Chu, *Journal of Environmental Management*, 2025, **392, 126798**.

45 T. Gao, T. Dai, N. Fan, Z. Han and X. Gao, *Journal of Environmental Management*, 2024, **363, 121314**.

46 Y. Liu, Z. Shu, Y. Ding, T. Yan, Q. Xu, L. Chen, A. Wang, M. Li, Y. Ma, J. Jiang and J. Wu, *Journal of Energy Storage*, 2025, **112, 115497**.

47 F. Kang, Y. Zhao, Y. Liu, J. Shen, W. Liu, W. Lv, T. Cao, H. Cao and Z. Sun, *ACS Sustainable Chemistry & Engineering*, 2025, **13**, 5985–5995.

48 H. Wang, Y. Liu, M. Jiang and Q. Zhang, *Journal of Energy Chemistry*, 2025, **101**, 437–452.

49 J.-Y. Lee, H. S. Yang, S. Hwang, S.-H. Jeong and J. Moon, *Journal of Energy Storage*, 2025, **121, 116576.**

50 A. Mondal, Y. Fu, W. Gao and C. C. Mi, *Batteries*, 2024, **10, 37.**

51 H. Rouhi, R. Serna-Guerrero and A. Santasalo-Aarnio, *Journal of Energy Storage*, 2022, **55, 105734.**

52 N. Garg, S. Pekkinen, E. Martínez González, R. Serna-Guerrero, P. Peljo and A. Santasalo-Aarnio, *Sustainable Energy & Fuels*, 2024, **8**, 2777–2788.

53 L. P. Yao, Q. Zeng, T. Qi and J. Li, *Journal of Cleaner Production*, 2020, **245, 118820.**

54 S. Kim, J. Bang, J. Yoo, Y. Shin, J. Bae, J. Jeong, K. Kim, P. Dong and K. Kwon, *Journal of Cleaner Production*, 2021, **294, 126329.**

55 L. Wu, F.-S. Zhang, K. He, Z.-Y. Zhang and C.-C. Zhang, *Journal of Cleaner Production*, 2022, **380, 135045.**

56 L. Wu, F.-S. Zhang, Z.-Y. Zhang and C.-C. Zhang, *Separation and Purification Technology*, 2023, **306, 122640.**

57 B. Li, T. Zhou, H. Sun, Q. Song, Y. Liu, J. Li and Z. Xu, *Journal of Hazardous Materials*, 2025, **486, 137036.**

58 Z. Zhu and X. Gao, *Advanced Sustainable Systems*, 2024, **9, 2400610.**

59 C. Yang, J.-l. Zhang, Q.-k. Jing, Y.-b. Liu, Y.-q. Chen and C.-y. Wang, *International Journal of Minerals, Metallurgy and Materials*, 2021, **28**, 1478–1487.

60 X.-c. Zhang, T. Zhang, N.-n. Liu, X.-d. Yin, X.-n. Wu, H.-l. Han, Q.-l. Wang and Y.-j. Zhang, *Engineering Failure Analysis*, 2023, **154, 107653.**

61 M. Colledani, L. Gentilini, E. Mossali and N. Picone, *CIRP Annals*, 2023, **72**, 17–20.

62 D. Weigl and D. Young, *Resources, Conservation and Recycling*, 2023, **192, 106936.**

63 H. Ali, H. A. Khan and M. Pecht, *Renewable and Sustainable Energy Reviews*, 2022, **168, 112809.**

64 G. Zhang, Y. He, H. Wang, Y. Feng, W. Xie and X. Zhu, *Journal of Cleaner Production*, 2019, **231**, 1418–1427.

65 C. Wang, R. Tao, P. Xing, J. Ni, Z. Cun, S. Liu, X. Wang, Y. Dai, Z. Long, Z. Sun and H. Li, *Resources, Conservation and Recycling*, 2025, **218, 108257.**

66 H. Qiu, C. Peschel, M. Winter, S. Nowak, J. Köthe and D. Goldmann, *Metals*, 2022, **12, 677.**

67 Z. Hu, J. Liu, T. Gan, D. Lu, Y. Wang and X. Zheng, *Separation and Purification Technology*, 2022, **297, 121486.**

68 K. Jia, G. Yang, Y. He, Z. Cao, J. Gao, H. Zhao, Z. Piao, J. Wang, A. M. Abdelkader, Z. Liang, R. V. Kumar, G. Zhou, S. Ding and K. Xi, *Advanced Materials*, 2024, 36, 2313273.

69 J. Li, P. Su, J. Li, Y. Liu, J. Wang, H. Khoshdast, Y. He and S. Nazari, *Journal of Power Sources*, 2024, **611, 234773.**

3

Cascaded Utilization

Zeyu Dong, Gaoyun Tan, Zihao Lu, Yunpeng Wen, Runxuan Chen, Xizhuo Chen, Yue Yang

3.1 The Hub of Resource Recycling and Economic Benefits

Retired batteries are precisely sorted through multidimensional testing (e.g., State of Health [SOH], internal resistance, and remaining capacity). Approximately 20–35% of batteries with satisfactory performance enter the cascaded utilization stream, while the remainder are sent for dismantling and recycling. This screening process significantly enhances resource value—cascaded utilization can increase the full lifecycle value of batteries by over 300%. For instance, when vehicle power batteries (with capacity decayed to 70–80%) are repurposed into energy storage systems (ESS), their residual value can reach 3 to 5 times that of metal recovery through dismantling. They can continue operating for 5 to 8 years, thereby delaying the entry of resources into high-energy-consumption recycling processes. From a technical standpoint, integrating electrochemical impedance spectroscopy (EIS) with machine learning algorithms is necessary to achieve real-time health assessment and to overcome engineering challenges such as module reconfiguration and heterogeneous system compatibility management. Tesla's Megapack ESS, which undergoes over 2,000 parameter tests to reconstruct its thermal management architecture, exemplifies the associated technical challenges and the focus on energy density per unit cost.

3.2 Resource Efficiency and Low-carbon Transition Cornerstone

Cascaded utilization mitigates the risk of critical metal supply shortages: 60% of the world's cobalt is produced in the Democratic Republic of the Congo, and lithium resources are concentrated in South American salt lakes and Australian mines. Each 1 GWh of secondary batteries can reduce the demand for 500 tons of lithium and

200 tons of cobalt, while also lowering carbon emissions by 3,000 tons.[1] From a full lifecycle perspective, cascaded utilization reduces the carbon footprint of batteries by 40%, given that the production process accounts for 30–40% of the total emissions of electric vehicles (EVs).[2] Economically, the procurement cost of secondary batteries is only 30–50% that of new batteries. China Tower has cumulatively deployed over 500,000 sets, saving 1.2 billion CNY annually. The "battery leasing model" of CATL and Tesla's Powerwall business model further drives the industry chain towards a "manufacturing and service" transformation.

3.3 Safety Compliance and Systemic Necessity

Safety assurance: Hidden defects in retired batteries (e.g., lithium plating and SEI layer thickening) may cause thermal runaway. Notably, 62% of energy storage fires in South Korea (2017–2019) were attributed to undetected flaws in secondary batteries. Implementing strict testing standards (e.g., internal resistance $\leq 3\ m\Omega$, self-discharge rate ≤ 5 mV/day per cell) can reduce system failure rates from 12 to 1% (as verified by UL 1974). Advanced technologies, including ultrasonic electrode integrity scanning and digital twin simulation for cell balancing, help ensure that the cycle life of secondary systems exceeds 80% of that of new batteries.

Global compliance: The EU's "New Battery Regulation" (2023) mandates "digital passports" and CE certification testing; China's "Recycling and Utilization Measures" stipulate 16 mandatory testing thresholds. Cascaded utilization reduces the carbon footprint per kWh of batteries by 2.8 kg CO_2, making it a key strategy for complying with regulations like the EU's "Carbon Border Adjustment Mechanism (CBAM)."[3,4]

Rigid demand: The global volume of retired batteries is projected to surge from 520,000 tons in 2023 to 12 million tons by 2030.[5] Cascaded utilization delays the dismantling of over 50% of retired batteries by 5 to 8 years, providing a buffer period for upgrading metallurgical capacity and reducing the energy consumption of resource recycling by 65%. Meanwhile, secondary batteries, with a cost advantage of 0.4 to 0.6 CNY/Wh, support the new energy system. In 2023, China's secondary energy storage installations reached 2.4 GWh, and a State Grid pilot project achieved a 30% cost reduction. In the low-speed EV sector, the proportion of secondary batteries exceeds 60%.

After long-term cycling, the performance of retired lithium-ion batteries (LIBs) degrades significantly, and their consistency is typically poorer than that of batteries in new energy vehicle systems. Battery consistency refers to the convergence of important characteristic parameters within a battery pack. Ideally, for multiple cells in the same pack, each parameter should fall within a narrow range, indicating good consistency.[6,7] Considering temporal factors, consistency requires the characteristic parameters of all cells to remain aligned throughout their entire lifecycle, accounting for inconsistencies in capacity attenuation, internal resistance growth, and aging rates. Pursuing consistency aims not only to maximize the current performance of the battery pack (including maximum power, current, and available capacity) but also to extend its service life. Consequently, prior to secondary use, retired batteries require classification and regrouping—specifically, grouping cells with similar performance into the same category and regrouping different types of batteries for different secondary use scenarios. Retired LIBs may be in an accelerated degradation phase. The inconsistency among individual cells can compound, seriously affecting safety.[8] Therefore, improving sorting accuracy is crucial for enhancing the safety and economy of secondary use.

3.4 Sorting of Retired LIBs

The most fundamental and core issue in sorting and regrouping is defining clear parameter-based grading criteria for retired LIBs. Numerous parameters could be applied to describe the residual value and health status of retired batteries, such as battery appearance, internal resistance, voltage, SOH, remaining life, etc.[9, 10] Theoretically, the performance degradation was triggered in retired LIBs, always through the fading of molecular-scale structure or material morphology (Figure 3.1a).[11] Thus, compared to other manners, using micro-scale sorting indices was regarded as the most direct and accurate approach. However, detecting these

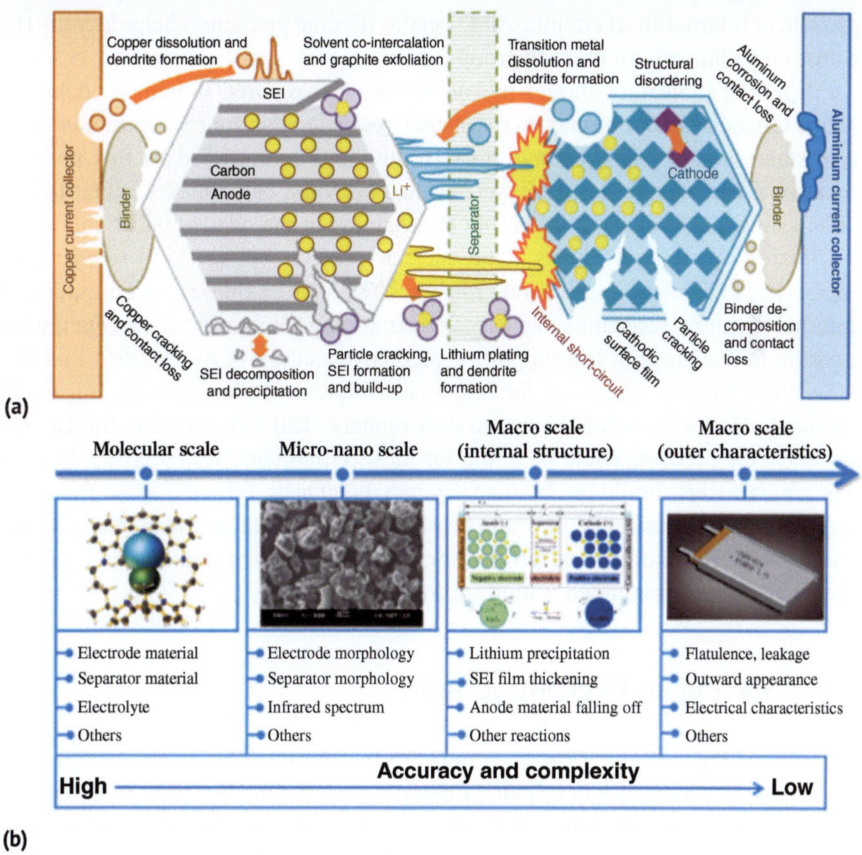

Figure 3.1 (a) Main failure mechanisms of LIBs. *Source*: [11] / Elsevier / CC BY 4.0; (b) screening criteria at different scales. *Source*: [12] / Elsevier / CC BY 4.0.

micro-scale indices is cumbersome, always requiring high-precision equipment. Thus, it is not economically or efficiently feasible for large-scale sorting. Therefore, macro-scale features are typically used as sorting indices. Some battery indicators were always used, including battery appearance, capacity, remaining life, internal resistance, impedance, and charge-discharge curve characteristics. However, limited by the inaccurate traits of these macro-scale data, they hardly reflect the internal state of retired batteries, particularly their degradation mechanisms. As the complexity of sorting indicators, detection accuracy was always decreasing (Figure 3.1b). As known, for retired LIBs, a typical sorting and reorganization process was composed of state assessment, sorting, classification, and reorganization. About state assessment, they could determine the efficiency and quality of the entire sorting and regrouping process.

To enable large-scale cascaded utilization of retired LIBs, a multidimensional sorting system is essential for a comprehensive state evaluation.[13] Relying on a single parameter, especially only about capacity alone, is always insufficient, and it may group batteries with similar capacity but vastly different internal resistance, leading to suboptimal performance. Safety was regarded as the foremost priority, always requiring consideration of failure-related factors, like lithium plating, SEI growth, or internal short circuits. Additionally, lifetime prediction helps identify the future degradation path for each group.

Integrating indicators should be carried out across three timelines, including characteristic degradation modes (past state), capacity and internal resistance (present state), and degradation trajectory (future state). Supported by this multidimensional assessment, a full lifecycle evaluation could be applied, facilitating the enhancement of both safety and economic viability in cascaded applications.

Utilizing operational history data, such as current, voltage, and temperature logs, is key to efficient and accurate sorting.[16] This data enables rapid estimation of internal reactions, resistance, capacity, and remaining life, significantly reducing the need for manual testing and supporting predictive analysis of future performance.

Sorting efficiency is critical for large-scale implementation, directly impacting economic feasibility. While historical data renders ideal rapid sorting, the lacking of usable data records could be found for most retired batteries currently. In such cases, parameters should be obtained through direct tests, bringing about the selection of quickly measurable indicators essential. Traditional methods that measure each parameter separately are too slow for industrial-scale sorting, underscoring the need for efficient, integrated testing approaches.[14, 15]

3.5 Sorting and Reconfiguration

After extracting multidimensional parameters to determine the state of retired LIBs, the critical challenge lies in utilizing these parameters for effective sorting. Battery sorting and reconfiguration constitute a multidimensional classification or clustering problem—a classic challenge in machine learning. Thus, common clustering algorithms like K-means clustering and support vector machine (SVM) regression analysis serve as key technologies for classifying and reconfiguring retired LIBs.

K-means clustering is a prevalent technique. Its workflow involves initialization, sample distance calculation, centroid update, and iteration. Before clustering, initial cluster centers (centroids) are randomly selected based on a preset cluster count. Distances between samples and centroids are computed, assigning each sample to the nearest cluster. Centroids are recalculated, and the process repeats until centroids stabilize or iterations conclude (Figure 3.2a). Methods include hard clustering (each cell belongs to one cluster; Figure 3.2b) and soft clustering (cells may belong to multiple clusters). K-means groups batteries with similar performance (e.g., capacity, internal resistance, voltage, and SOH) to enhance consistency after reconfiguration. For instance, Yin et al.[17] proposed a two-stage echelon utilization method combining static and dynamic characteristics (Figure 3.2c, d). Stage 1 preliminarily

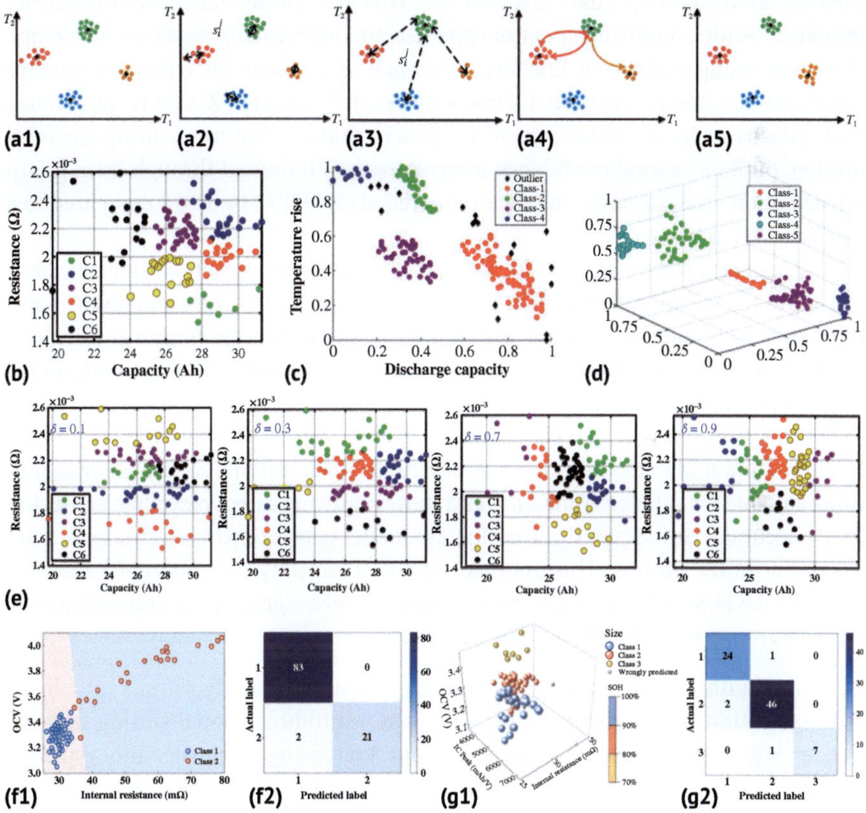

Figure 3.2 (a) Step-by-step process of K-means clustering algorithm: (a1) initial classification, (a2) intracluster distance calculation, (a3) intercluster distance calculation, (a4) reconfiguration, (a5) outcome. *Source*: [18] / with permission of IEEE; (b) typical K-means classification result. *Source*: [12] / Elsevier / CC BY 4.0; (c) K-means using static indicators (capacity, temperature). *Source*: [17] / with permission of Elsevier; (d) K-means using dynamic indicators (charge/discharge curves). *Source*: [17] / with permission of Elsevier; (e) K-means results under varying weight coefficients. *Source*: [12] / Elsevier / CC BY 4.0; (f) SVM binary classification: (f1) boundary, (f2) confusion matrix. *Source*: [19] / with permission of Elsevier; (g) SVM multidimensional classification: (g1) 3D distribution, (g2) confusion matrix. *Source*: [19] / with permission of Elsevier.

screens capacity and temperature rise to exclude faulty or hazardous batteries. Stage 2 uses dynamic features from discharge voltage curves with K-means for refined sorting. Post-sorting, standard deviations for capacity and temperature rise were below 0.06 and 0.41, respectively, indicating high intragroup consistency; low voltage drop and plateau deviations (<0.025 and <0.017) confirmed similar health states.

Conventional K-means can be sensitive to initial centroids and less adaptable to high-dimensional data, necessitating improvements. Dynamic feature weighting—assigning weights via a coefficient matrix—enhances adaptability to battery characteristics (Figure 3.2e). Another optimization is secondary clustering, subdividing initial clusters based on engineering constraints (e.g., cluster size limits) to boost practicality.

K-means excels in rapidly handling unlabeled data, suiting bulk sorting; optimized versions handle complex data distributions. However, preset cluster counts heavily influence results, requiring operator expertise and introducing potential instability. Moreover, using traditional Euclidean distance as a metric for battery parameter differences is sometimes debated. Thus, Chen et al.[20] validated K-means consistency via t-tests, using incremental capacity (IC) curve peaks as features. During reconfiguration, outliers beyond confidence intervals were eliminated through intra-group t-distribution analysis, reducing mean squared error (MSE) by >60% compared to a baseline MSE of 0.04.

SVM regression analysis, unlike K-means, requires no preset parameters, handles nonlinear relationships, offers clear classification boundaries, and resists noise interference. It finds a function minimizing prediction error within a tolerance while maintaining model simplicity. SVM regression excels in fine-grained, multidimensional classification and multilevel sorting for high-precision scenarios. Wu et al.[19] developed a two-step classification model using SVMs based on optimized IC curves under high charging rates. A binary classifier initially sorted batteries into those suitable for echelon utilization and those requiring direct recycling, achieving 98.1% accuracy. Subsequently, a multiclass model categorized batteries with SOH >70% into three groups. This approach reduced capacity and internal resistance standard deviations by up to 65 and 83%, respectively, significantly enhancing consistency among retired LIBs (Figure 3.2f, g).

Technical analysis of sorting and reconfiguration: Most algorithms suit binary classification of two-dimensional (2D) data. Multidimensional sorting necessitates building multidimensional classifiers. Increasing dimensionality typically elevates complexity and reduces efficiency. Therefore, application scenarios should be considered to minimize classification dimensions, reducing complexity while enhancing accuracy and efficiency. For instance, side reaction characteristics correlate with safety, while lifespan, internal resistance, and capacity relate to functionality. Safety is paramount. Primary sorting could use side reaction characteristics (a 1D problem), grouping batteries with similar safety profiles. Secondary sorting (a 3D problem) could then apply lifespan trajectory, capacity, and internal resistance to safety-matched groups, incorporating application scenarios: energy-oriented scenarios are classified by capacity and remaining lifespan; power-oriented scenarios by internal resistance and remaining lifespan (simplifying to 2D). Hybrid scenarios require capacity, internal resistance, and remaining lifespan. Accordingly, Lai et al.[12] proposed a multidimensional, hierarchical method constrained by application scenarios (Figure 3.3), executing two-level sorting: Level 1 by side reaction characteristics; Level 2 sorts safety-matched batteries by lifespan trajectory, capacity, and internal resistance.

Development requirements for sorting and reconfiguration technologies: Advancements must address four core needs: high-precision sorting, low-loss reconfiguration, scenario adaptability, and industrial synergy.

Sorting technology: Requires intelligent multimodal systems integrating AI visual inspection (appearance), spectral analysis (material composition), and dynamic electrochemical testing (voltage/resistance/cycle count) for multiparameter assessment, enhancing efficiency across cell formats (cylindrical/prismatic/pouch). Develop privacy-preserving federated learning models predicting SOH via minimal cycles to prevent data leakage. Implement RFID/blockchain lifecycle traceability to identify materials (e.g., LFP/NCM), auto-match protocols, and increase cross-regional compatibility by 40%.

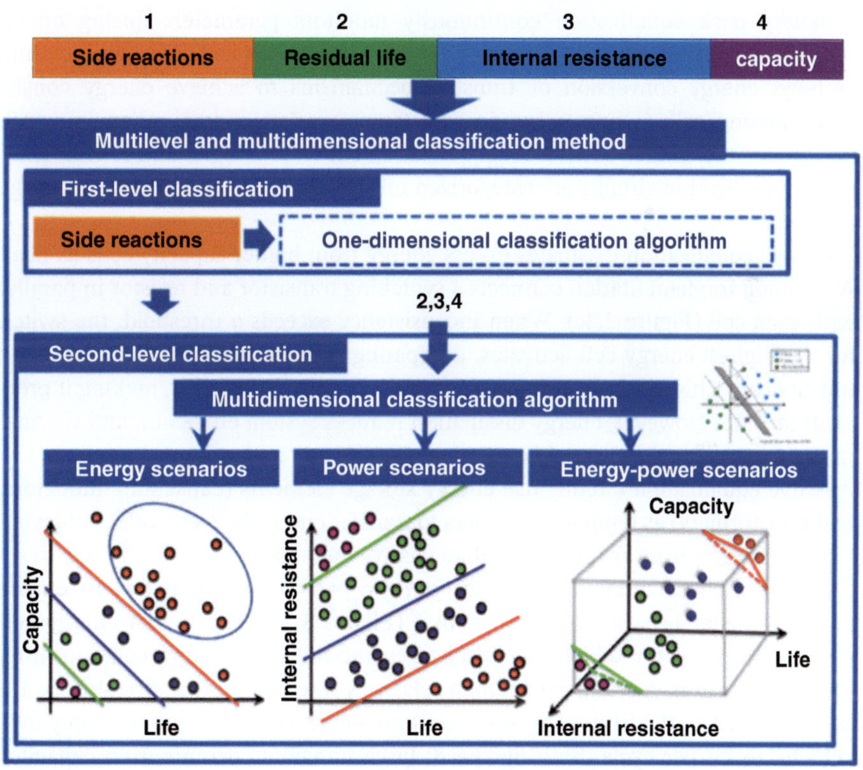

Figure 3.3 Multilevel and multidimensional classification method constrained by echelon utilization scenarios. *Source*: [12] / with permission of Elsevier.

Reconfiguration technology: Must resolve low-damage disassembly and cost-efficiency. Develop modular equipment using laser precision tab cutting and flexible shell delamination for 80% of mainstream formats. Adopt "live reconfiguration" (e.g., reusing functional modules without full disassembly, as in GM's 20 GWh project) to reduce disassembly costs by 30%. Implement scenario-driven hierarchical reconfiguration to lower new battery procurement costs by 60%. Future goals include zero-loss disassembly, full-scenario adaptive reconfiguration, and closed-loop management, aiming to slash echelon utilization costs by 50% and transform retired batteries into valuable strategic assets.

3.6 Equalization Technology

During echelon utilization, retired power batteries are typically reconfigured into packs via series-parallel connections to meet specific voltage, power, and capacity requirements. However, inconsistencies in capacity and internal resistance among cells cause variations in voltage change rates during charging/discharging, negatively impacting overall pack performance (Figure 3.4a, b). Equalization technology mitigates these effects.[21]

Battery pack equalization continuously monitors parameters during operation to prevent overcharging/overdischarging. It assesses cell inconsistency and employs energy conversion or transfer mechanisms to achieve energy consistency among cells,[22] preventing inconsistency escalation and enhancing pack performance. Research focuses on equalization circuit topologies and control strategies. Equalization circuits are categorized into passive and active based on energy handling.

Passive equalization dissipates excess energy from higher-capacity cells as heat. A common implementation connects a switching transistor and resistor in parallel with each cell (Figure 3.4c). When inconsistency exceeds a threshold, the switch for the highest-energy cell activates, dissipating surplus energy as heat, reducing imbalance.[23] This topology is simple, easy to control, and low-cost, making it prevalent in BMS. However, energy dissipation reduces system efficiency, and thermal management is critical to avoid compromising lifespan and safety.

Active equalization circuits use energy storage elements (capacitors, inductors, and transformers) as temporary carriers to transfer energy between cells, achieving higher energy utilization efficiency than passive methods.

Capacitor-based circuits (single or multi-capacitor) use switches to enable energy transfer between any two cells[24] (Figure 3.4f). The higher-energy cell charges the capacitor, which then discharges into the lower-energy cell, repeating until equilibrium. Advantages include relatively low complexity, straightforward control, good scalability, compact size, and lower cost. Limitations: only two cells balanced per cycle, limiting speed; large transient currents from capacitor charging/discharging stress components; balancing current depends on voltage difference, slowing process for small differences. Shang et al.[25] proposed an optimized mesh-structured switched-capacitor equalizer (MSSCE) offering the

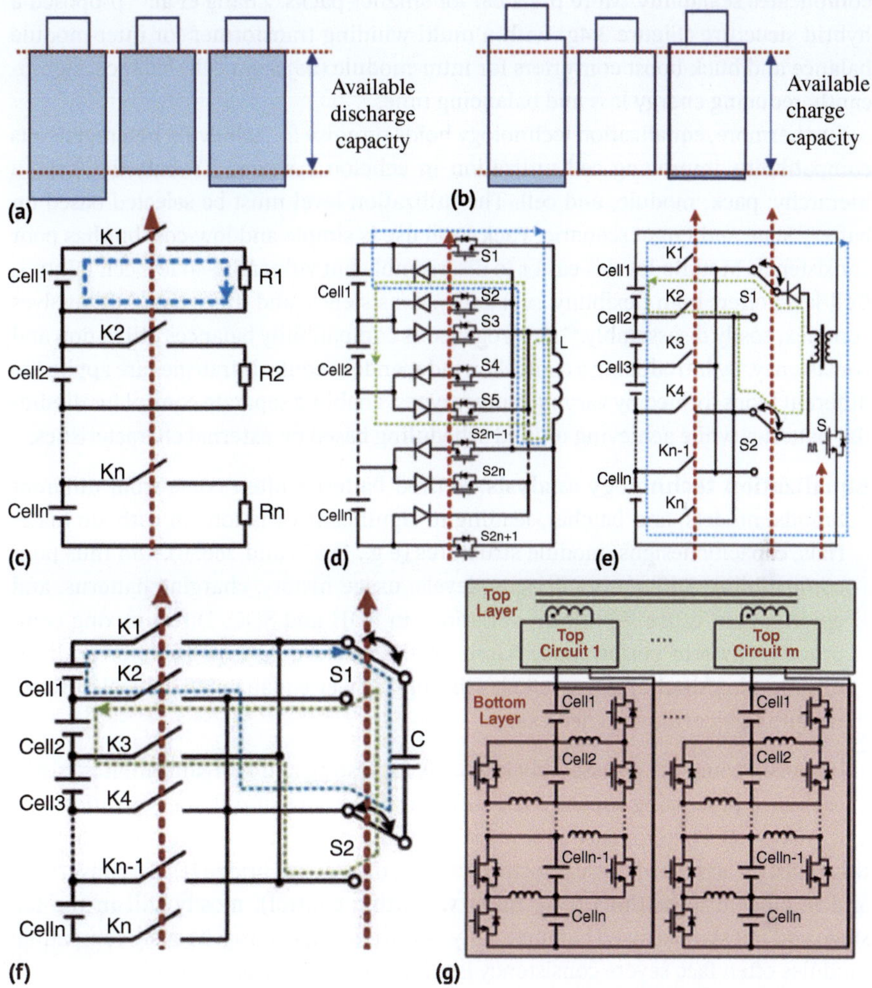

Figure 3.4 Impact of inconsistency: (a) usable discharge capacity, (b) usable charge capacity, (c) passive equalization circuit, (d) inductor-based active equalization, (e) transformer-based active equalization, (f) capacitor-based active equalization, and (g) hybrid equalization circuit. *Source*: [22] / with permission of John Wiley & Sons.

shortest paths between any cells, significantly enhancing speed and efficiency (91.7% improvement demonstrated) with robustness against imbalance.

Inductor-based circuits (single or multi-inductor) use an inductor and switching matrix for energy transfer[26] (Figure 3.4d). The equalization rate is not affected by voltage difference magnitude, improving speed versus capacitors. However, using only one storage element restricts simultaneous balancing to two cells, increasing time for large packs and limiting the rate.

Transformer-based circuits use an isolated structure, often based on forward/flyback converters, using the transformer itself for energy transfer between cells or between cells and the pack[27] (Figure 3.4e). Simple and easy to control, but speed is moderate, and requiring high consistency in multiple secondary windings complicates scalability. More practical for smaller packs. Zhang et al.[28] proposed a hybrid structure (Figure 3.4g) with a multi-winding transformer for inter-module balance and buck-boost converters for intra-module (adjacent cell) balance, significantly reducing energy loss and balancing time.

Furthermore, equalization technology holds promise for achieving heterogeneous compatibility, improving cell utilization in echelon systems. EV batteries have a hierarchy: pack, module, and cell. The utilization level must be selected based on battery state and target scenario. Pack-level use is simple and low-cost but has poor consistency. Module-level is easier to disassemble but vulnerable to few cell failures. Cell-level offers high flexibility, scalability, consistency, and utilization but involves complex, costly disassembly.[29] Heterogeneous compatibility balances utilization and consistency. By introducing a controller, independent control strategies are applied to different loops formed by varying battery types, enabling separate control for dissimilar batteries while achieving unified scheduling based on external characteristics.

Equalization technology analysis: Retired batteries often come from different brands, models, and batches, leading to significant variations in cathode materials, capacity designs, module structures (e.g., 18650 and 26650), and thus poor compatibility. Differences in aging levels, usage history, charging patterns, and environment cause significant variations in SOH and SOC. Direct mixing compromises system performance based on the weakest battery. Therefore, adjusting charging/discharging strategies through power equalization control or fuzzy weighting algorithms is necessary.[30]

However, complete disassembly into cells for testing and reorganization is costly. To enhance processing capacity and efficiency, batteries of different specifications and grades need heterogeneous compatibility and reassembly before use. Achieving this relies on advanced power electronics (e.g., dual active bridge [DAB] converters) and intelligent algorithms (e.g., fuzzy weighting control), mostly still in the lab stage and not standardized commercially. Additionally, heterogeneously compatible modules often face severe consistency issues, limiting echelon utilization.

Development needs for equalization technology: Reassembling batteries of the same level improves processing capacity and efficiency. However, technologies for heterogeneous compatibility of different grades suffer from low efficiency and poor safety. Lack of unified standards leads to serious consistency problems in reassembled batteries. Further development is needed to improve compatibility efficiency for different grades and effectively control the consistency of reassembled modules, enabling efficient reorganization and expanding application scenarios.

3.7 Safety Management Technology

Echelon utilization involves reusing mildly retired batteries, key to efficient value-added use. However, diverse sources and significant characteristic variations increase safety risks upon direct grouping.[31] Thus, monitoring operating conditions, enhancing fault diagnosis and protection functions, conducting regular insulation tests and maintenance, and improving the reliability of protection devices (leakage, over-voltage, over-current) are necessary.[32]

The battery thermal management system (TMS) is core to safety. Most current TMS use forced air cooling, which has a simple structure (Figure 3.5a) and mature technology. Liquid cooling offers significantly higher heat transfer efficiency[33] but is more complex (Figure 3.5b) due to dense packing in ESS. Coolant leakage can easily cause short circuits, chain reactions, and safety accidents. To avoid leakage, Bamdezh et al.[34] proposed a hybrid system using cooling water channels and a foam-phase change material-aluminum composite around batteries, studying the effect of foam anisotropy.

Low-temperature heating technology is key for operation in cold environments. Current preheating/heating technologies use PTC (Figure 3.5c) or electric heating films to heat the coolant.[39] Some studies use high latent heat phase change materials to absorb waste heat and release it slowly when needed. Others use insulation materials like aerogels[40] to prevent heat loss. TMS design must consider battery aging patterns, performance, cost, space, efficiency, and safety, with differentiated optimization based on application scenario requirements.[36]

To address limitations of traditional TMS in thermal safety, improvements are needed in monitoring, early warning, prevention, control, and firefighting. First, the thermal runaway behavior characteristics and temperature thresholds of retired power LIBs must be clarified for scientific monitoring. However, most studies use new batteries, whose behavior differs significantly from retired ones, affecting accuracy (Figure 3.5d). Numerical simulation can effectively analyze key parameters of retired batteries (stability, safety, and heat generation). By constructing multidimensional heat generation and thermal runaway models integrating failure mechanisms and key parameters, safe operational limits and safety thresholds for echelon

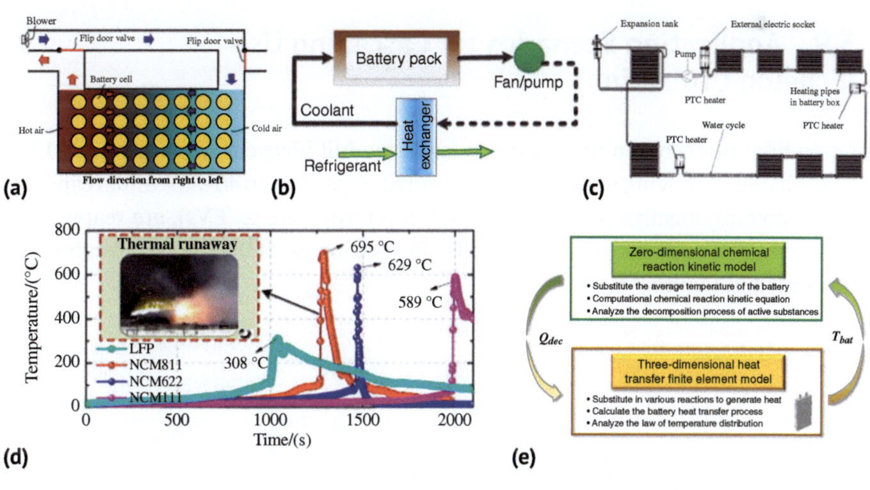

Figure 3.5 (a) Air cooling structure. *Source*: [35] / with permission of Elsevier; (b) liquid cooling structure. *Source*: [35] / with permission of Elsevier; (c) PTC heating. *Source*: [36] / with permission of Elsevier; (d) thermal runaway behavior in different models. *Source*: [37] / with permission of Elsevier; (e) thermal runaway propagation modeling framework [38].

utilization can be defined.[41] Li et al.[38] established a thermal runaway propagation model within a module based on heat transfer theory (Figure 3.5e), simplified dimensions, and modified the input source term based on a lumped model. They developed a temperature distribution calculation model consistent with cell-level results, usable for module temperature prediction and early warning.

Safety management technology analysis: Currently, retired LIBs are often used at scale without matching BMS; most BMS are designed for conventional new batteries. However, retired batteries have degraded performance due to aging.[15] Even after strict classification and reassembly, consistency in reused modules is lower than in new batches, leading to ineffective capacity utilization and potential safety risks, especially if batteries are in an accelerated aging phase.

3.8 Safety Management Development Needs: Development of Highly Compatible BMS Systems

The BMS is critical for balanced regulation and safety supervision. Most large-scale echelon systems currently use BMS designed for traditional new batteries, with principles and thresholds developed for them.

Batteries in echelon use are highly aged. Due to accelerated aging, their performance may fluctuate significantly. Reassembled batteries have poorer consistency, making them more susceptible to failures and safety issues. Therefore, exploring the dynamic state trajectory of retired batteries during reuse, investigating long-term parameter changes, and further clarifying threshold conditions are necessary to develop a BMS tailored for echelon utilization, ensuring compatibility and safety.

3.9 Application Scenarios for Cascading Use (Echelon Utilization)

Cascading use (echelon utilization) extends the full lifecycle value of retired LIBs. These batteries, having undergone capacity fade (typically to 40–80% initial capacity) and becoming unsuitable for high-power applications (e.g., EVs), are reapplied in fields with lower performance demands after testing, sorting, and repackaging,[42] as illustrated in Figure 3.6.

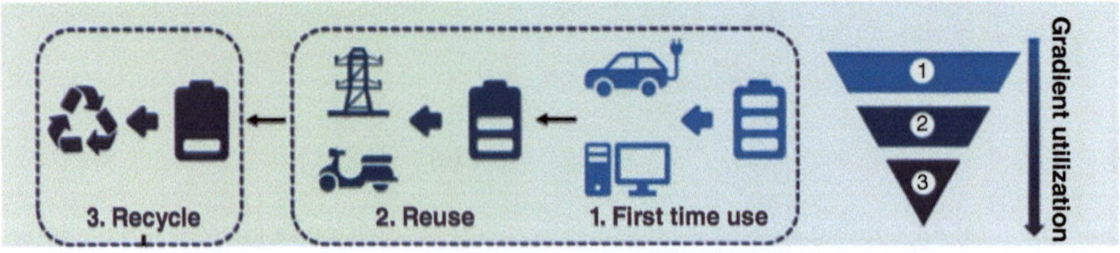

Figure 3.6 Schematic diagram of LIB echelon utilization pathways.

Table 3.1 Application scenarios for batteries of different capacity tiers.

Application scenarios	Capacity requirements	Cycle life (cycles)	Cost (CNY/kWh)
EVs	≥80%	>1,000	1,200–1,500
ESS	60–80%	500–800 cycles	600–800
Low-speed EVs	50–70%	300–500 cycles	400–600
Backup power systems	≥40%	200–300 cycles	300–500

After retirement from EVs, LIBs retain usable capacity. They can be repurposed through "downgraded application" for lower-intensity scenarios (e.g., ESS, backup power, and low-speed EVs) (Table 3.1), rather than being directly disassembled for recycling.[43] Echelon utilization extends service life, reduces energy consumption and carbon emissions, offering significant economic and environmental benefits.[44, 45] Globally, the echelon utilization market was valued at ~$8 billion in 2023 and is projected to grow to $65 billion by 2030 (35% CAGR), potentially accounting for 45% of the total battery recycling market. In China, installed capacity reached 5 GWh in 2023 (mainly ESS and LSEVs), forecast to exceed 15 GWh by 2025 (China Automotive Technology and Research Center).

1. **Energy storage systems**
   ESS application is a primary direction for echelon utilization.[46] ESS have lower performance requirements and tolerates some degradation. Retired batteries can be used in residential ESS to store surplus solar/wind power or in grid-level ESS for load balancing and enhancing stability/reliability. As the dominant market, ESS deployments include China Tower's installation of over 50 GWh of cascaded batteries across 500,000 base stations, replacing 3 million tons of lead-acid batteries and reducing annual electricity costs by CNY 1.2 billion—the world's largest echelon utilization project.
2. **Backup power systems**
   Backup power is another critical domain.[47] These systems prioritize instantaneous discharge capability and cycle life over energy density/capacity. Retired batteries are used in telecom base station backup power to ensure uninterrupted operation. Major operators (e.g., China Mobile and Vodafone) use cascaded batteries to replace diesel generators, achieving 90% carbon reduction. They also serve at UPS for data centers. Companies like Google and Amazon use repurposed batteries in some data centers, achieving 92% round-trip efficiency at 50% lower cost than new batteries.
3. **Low-speed EVs (LSEVs)**
   Echelon utilization in LSEVs—including e-bikes, e-scooters, and electric forklifts—is a key growth area.[48] While unsuitable for EVs, retired batteries are ideal for short-distance mobility (e.g., shared e-bikes) and logistics (e.g., warehouse forklifts). Studies confirm repurposed batteries outperform lead-acid in electric forklifts and low-speed utility vehicles,[49] offering significant economic

Table 3.2 Four core advantages of echelon utilization.

Advantages	Specific manifestations	Data supports
Advantages	Reduced demand for critical mineral extraction	1 GWh of cascaded-use batteries reduces the extraction of approximately 500 metric tons of lithium ore and 200 metric tons of cobalt ore. *(Source: Benchmark Minerals Intelligence)*
Pronounced economic benefits	Cascaded battery costs at 30–50% of the same-capacity new batteries, reducing ESS costs by 40%	China Tower achieves annual cost savings exceeding CNY 1 billion through cascaded battery deployment. *(Source: 2023 Financial Report)*
Environmental friendliness	Delays battery dismantling, cutting full lifecycle carbon emissions by 30–50%	Tesla Powerwall systems utilizing cascaded batteries demonstrate 35% lower carbon emissions compared to new battery systems. *(Source: Tesla 2023 Impact Report)*
Extended battery service life	Extends total battery service life by 5–10 years, enabling dual-use application	BMW's Leipzig storage station has maintained stable operation for 4 years using cascaded batteries, demonstrating an annual capacity fade rate of only 5%

advantages. Ola Electric (India) reduced range costs by 40%. Chinese manufacturers like Yadea offer e-bikes with repurposed batteries at 30% lower prices. JD Logistics saved over CNY 10 million annually in electricity costs with battery-powered forklifts, extending battery replacement cycles by 30%.

4. **Emerging applications**
Retired batteries are also trialed in niche applications:[50] streetlight energy storage: Shenzhen pilots reduced nighttime power costs by 30%, marine auxiliary power: Norwegian ferries cut fuel consumption by 15%. In summary, echelon utilization maximizes residual value through "downgraded, not discarded" reuse, with broad applications in ESS, backup power, and LSEVs. Its core advantages (Table 3.2) include extended service life, reduced energy consumption, and lower carbon emissions, delivering substantial economic and environmental benefits. With advancements in testing, BMS, and policy support, echelon utilization will become a critical growth vector for the battery industry.

3.10 Chapter Summary

Echelon utilization enables high-value extraction of residual potential from multisource retired LIBs while ensuring long-term operational safety. It represents repetitive reuse after mild first-life retirement, serving as the critical pathway for value-added retirement management. However, heterogeneous sources with substantial performance variability heighten safety risks upon direct repackaging. System capacity is constrained by the lowest-capacity unit, and electrochemical divergence reduces residual capacity utilization efficiency, impeding high-grade applications.

Crucially, unstable electrochemical characteristics in retired batteries elevate thermal runaway risks during intensive use. Therefore, developing residual capacity optimization protocols within safety governance frameworks is essential for maximizing lifecycle value—constituting an indispensable component of closed-loop recycling value chains.

The ideal echelon utilization framework requires:

1. Standardized sorting criteria tailored to application scenarios (power/capacity/internal resistance thresholds).
2. Intelligent repackaging systems ensuring module consistency through optimized clustering algorithms.
3. Specialized BMS enabling heterogeneous compatibility and real-time safety monitoring.
4. Application diversification into grid storage, renewable integration, and microgrids.

References

1 H. Chen, T. Zhang, Q. Gao, Z. Han, Y. Jin, L. Li, K. Yang, Y. Xu, X. Liu, X. Xu and S. Wang, *Journal of Cleaner Production*, 2022, **379**, 134583.

2 S. Guan, H. Tong, Z. He, Z. Dong and M. Gao, *Journal of Energy Storage*, 2025, **132**, 117786

3 C. Zhang, Y.-X. Tian and M.-H. Han, *Journal of Cleaner Production*, 2022, **375**, 134060

4 X. Gu, H. Bai, X. Cui, J. Zhu, W. Zhuang, Z. Li, X. Hu and Z. Song, *Renewable and Sustainable Energy Reviews*, 2024, **192**, 114191.

5 IEA, Global EV Outlook 2023: catching up with climate ambitions, 2023. https://www.iea.org/reports/global-ev-outlook-2023.

6 L. Chang, C. Ma, Y. Zhang, H. Li and L. Xiao, *Journal of Energy Storage*, 2022, **55**,105539

7 M. Ecker, N. Nieto, S. Käbitz, J. Schmalstieg, H. Blanke, A. Warnecke and D. U. Sauer, *Journal of Power Sources*, 2014, **248**, 839–851.

8 B. Wang, Y. Dong, Z. Zeng, H. Lei, Y. Yang and P. Ge, *Journal of Power Sources*, 2024, **621**, 235299.

9 M. F. Roslan, P. R. Satpathy, T. Prasankumar, V. K. Ramachandaramurthy, M. Mansor and S. L. Walker, *Journal of Energy Storage*, 2025, **123**, 116808.

10 R. Pan, X. Xiao, J. Fan, M. Tan, X. Luo, J. Tian, D. Yang and J. Xu, *Journal of Energy Storage*, 2024, **99**, 113387.

11 C. R. Birkl, M. R. Roberts, E. McTurk, P. G. Bruce and D. A. Howey, *Journal of Power Sources*, 2017, **341**, 373–386.
12 X. Lai, Y. Huang, C. Deng, H. Gu, X. Han, Y. Zheng and M. Ouyang, *Renewable and Sustainable Energy Reviews*, 2021, **146**, 111162.
13 N. Wang, A. Garg, S. Su, J. Mou, L. Gao and W. Li, *Batteries*, 2022, **8**, 080096.
14 X. Xu, J. Mi, M. Fan, K. Yang, H. Wang, J. Liu and H. Yan, *Journal of Cleaner Production*, 2019, **213**, 1080–1086.
15 Z. Wei, X. Han and J. Li, *Journal of Energy Storage*, 2022, **51**, 104428.
16 X. Lai, Q. Chen, X. Tang, Y. Zhou, F. Gao, Y. Guo, R. Bhagat and Y. Zheng, *eTransportation*, 2022, **12**, 100169.
17 H. Yin, Y. Li, Y. Kang and C. Zhang, *Journal of Energy Storage*, 2023, **64**, 107178.
18 M. Wu, L. Wang, W. Xia and J. Wu, *IEEE Transactions on Transportation Electrification*, 2025, **11**, 1939–1948.
19 Z. Wu, Z. Zhao, Z. Fang, H. Bao, P. Liu, X. Su, Y. Zhang, K. Jin, H. Li and Q. Wang, *Journal of Power Sources*, 2025, **654**, 237845.
20 Z. Chen, Y. Deng, H. Li and W. Liu, *Journal of Energy Storage*, 2022, **56**, 105917.
21 J. Sun, C. Tang, X. Li, T. Wang, T. Jiang, Y. Tang, S. Chen, S. Qiu and C. Zhu, *eTransportation*, 2022, **11**, 100149.
22 Y. Hua, S. Zhou, H. Cui, X. Liu, C. Zhang, X. Xu, H. Ling and S. Yang, *International Journal of Energy Research*, 2020, **44**, 11059–11087.
23 F. Peng, H. Wang and L. Yu, *IEEE Transactions on Industry Applications*, 2019, **55**, 4053–4063.
24 L. Fang, J.-q. Li and W. Shi, *Energy Procedia*, 2017, **105**, 2725–2732.
25 Y. Shang, Q. Zhang, N. Cui, B. Duan and C. Zhang, *IEEE Transactions on Transportation Electrification*, 2019, **5**, 252–261.
26 D. D. Quinn and T. T. Hartley, *Journal of Power Sources*, 2013, **240**, 26–32.
27 W. Huang and J. A. Abu Qahouq, *IEEE Transactions on Industrial Electronics*, 2015, **62**, 2764–2776.
28 Z. Zhang, H. Gui, D.-J. Gu, Y. Yang and X. Ren, *IEEE Transactions on Power Electronics*, 2017, **32**, 2757–2768.
29 W. Li, Y. Peng, Y. Zhu, D. T. Pham, A. Y. C. Nee and S. K. Ong, *Robotics and Computer-Integrated Manufacturing*, 2024, **89**, 102758.
30 C. Ma, L. Chang, N. Cui, B. Duan, Y. Zhang and Z. Yu, *Energy*, 2022, **257**, 124692.
31 Q. Dong, S. Liang, J. Li, H. C. Kim, W. Shen and T. J. Wallington, *iScience*, 2023, **26**, 107195.
32 D. Kong, H. Lv, P. Ping and G. Wang, *Journal of Energy Storage*, 2023, **64**, 107073.
33 D. Chen, J. Jiang, G.-H. Kim, C. Yang and A. Pesaran, *Applied Thermal Engineering*, 2016, **94**, 846–854.
34 M. A. Bamdezh, G. R. Molaeimanesh and S. Zanganeh, *Journal of Energy Storage*, 2020, **32**, 101778.
35 R. Mahamud and C. Park, *Journal of Power Sources*, 2011, **196**, 5685–5696.
36 D. Shao, L. Hu, J. Zhang, R. Hu, G. Zhang, L. Jiang, X. Wang and Y. Wen, *International Journal of Electrochemical Science*, 2024, **19**, 100817.
37 Z. Wang, J. Yuan, X. Zhu, H. Wang, L. Huang, Y. Wang and S. Xu, *Journal of Energy Chemistry*, 2021, **55**, 484–498.

38 X. Li, F. Zhou, S. Gao, J. Zhao, D. Wang and H. Yin, *Waste Management*, 2022, **153**, 347–354.

39 B. Xu, J. Lee, D. Kwon, L. Kong and M. Pecht, *Renewable and Sustainable Energy Reviews*, 2021, **150**, 111437.

40 S. Wilke, B. Schweitzer, S. Khateeb and S. Al-Hallaj, *Journal of Power Sources*, 2017, **340**, 51–59.

41 D. Gao, Y. Du, Y. Cheng and Q. Yang, *Engineering Applications of Artificial Intelligence*, 2024, **132**, 107919.

42 R. Sathre, C. D. Scown, O. Kavvada and T. P. Hendrickson, *Journal of Power Sources*, 2015, **288**, 82–91.

43 S. Sun, C. Jin, W. He, G. Li, H. Zhu and J. Huang, *Science of the Total Environment*, 2021, **776**, 145913.

44 S. Lei, W. Sun and Y. Yang, *Environmental Science & Technology*, 2024, **58**, 3609–3628.

45 H. Zhang, J. Huang, R. Hu, D. Zhou, H. u. R. Khan and C. Ma, *Energy*, 2020, **206**, 118178.

46 Q. Wei, Y. Wu, S. Li, R. Chen, J. Ding and C. Zhang, *Science of the Total Environment*, 2023, **866**, 161380.

47 Y. Nie, Y. Wang, L. Li and H. Liao, *International Journal of Environmental Research and Public Health*, 2023, **20**, 054346.

48 M. H. S. M. Haram, J. W. Lee, G. Ramasamy, E. E. Ngu, S. P. Thiagarajah and Y. H. Lee, *Alexandria Engineering Journal*, 2021, **60**, 4517–4536.

49 D. G. Hanjiro Ambrose, Alexander Gershenson and Daniel Kammen, *Environmental Research Letters*, 2014, **9**, 094004.

50 Y. Hua, S. Zhou, Y. Huang, X. Liu, H. Ling, X. Zhou, C. Zhang and S. Yang, *Journal of Power Sources*, 2020, **478**, 228753.

4
Metallurgical Extraction Techniques

Xuejing Qiu, Peixiang Gao, Yimeng Zhang, Lingling Xie, Limin Zhu, Xiaoyu Cao

4.1 Pyrometallurgical Extraction Techniques

4.1.1 Application of Pyrometallurgical Recycling in the Lithium Iron Phosphate ($LiFePO_4$, LFP)

Pyrometallurgy, as a typical high-temperature metallurgical process, faces challenges in its application to LFP, primarily due to the difficulty in breaking down the stable material structure of LFP and the complexity of element separation. From the perspective of the recycling process, the pyrometallurgical method utilizes crushed black powder or electrode sheets as raw materials. Through high-temperature roasting (800–1,200 °C), it directly decomposes carbon-based binders (e.g., polyvinylidene fluoride [PVDF]), graphite anodes, and residual electrolytes (e.g., $LiPF_6$) in the battery, thereby achieving thermal decomposition and removal of organic components. Meanwhile, the LFP cathode undergoes decomposition in a reducing atmosphere, releasing Fe, P, and Li elements. Among these, Fe and P form an Fe-P alloy (or Fe_2P phase) through high-temperature slagging, which can be utilized as a steelmaking auxiliary material or as a raw material for phosphorus fertilizer. In contrast, lithium (Li), due to its high-temperature volatility, is enriched in the form of lithium oxide (Li_2O) in the flue gas ash or accumulates in the ash slag collected from the flue gas purification system. When LFP is subjected to direct high-temperature smelting, the proportion of lithium that volatilizes and is lost can reach as high as 30 to 40%. The residual Li_2O tends to react with silica present in the slag (often derived from the crucible's refractory material or impurities), forming insoluble Li_2SiO_3 and resulting in the so-called "lithium lock" effect. To overcome this issue, subsequent strong acid leaching (e.g., using a sulfuric acid–hydrogen peroxide [H_2O_2] system) is required to dissolve and recover the lithium. Furthermore, as the primary product of pyrometallurgical processes, the Fe-P alloy is sold at only 3,000 to 4,000 CNY/ton, whereas the smelting and processing cost for each ton of spent LFP material reaches as high as 6,000 to 8,000 CNY/ton. If the project relies solely on revenue from the Fe-P alloy, it is likely to incur significant losses. Finally, the high-temperature decomposition of the PVDF binder and $LiPF_6$ electrolyte salt can release highly toxic pollutants, such as hydrogen fluoride (HF) and fluorine-containing particulate matter. Thus, the pyrometallurgical production line must be equipped with a secondary off-gas purification system to achieve deep purification.[1, 2]

The equipment investment for this type of complex defluorination unit accounts for 25 to 30% of the total production line cost, and the continuous consumption of alkaline reagents during operation further compresses the profit margin. In the industry, relevant companies have attempted to improve the pyrometallurgical process. For example, the "directional calcination + acid leaching method" developed by the US ReCell Center can increase the overall lithium recovery rate to 65%, but the cost remains 15% higher than the market price of recycled lithium, and the process has not yet achieved commercial viability.

Relevant researchers have individually investigated pyrometallurgical processes. Georgi-Maschler et al. developed a selective pyrometallurgical treatment method using an electric arc furnace, which can partially convert the materials from spent lithium-ion batteries (LIBs) into lithium concentrate and enable lithium recovery in the form of lithium carbonate (Li_2CO_3) through hydrometallurgical steps. In addition, other material fractions, such as the iron fraction, aluminum fraction, and copper fraction, can be obtained for further processing. However, this method is associated with high energy consumption and relatively low metal recovery rates and purity.[3] Li et al. employed NaOH as an oxidant to disrupt the stable olivine structure of LFP during the calcination process, enabling low-temperature pyrometallurgical recovery of LFP at 150 °C. This approach effectively addresses the issues of high energy consumption and low recovery efficiency commonly associated with conventional pyrometallurgical recovery methods.[4] Yin et al. utilized Na_2CO_3 as an activator to break the chemical bonds in LFP, facilitating the carbothermal reduction of LFP into Fe, $NaLi_2PO_4$, and $LiNa_5(PO_4)_2$. Following magnetic separation, the lithium recovery rate reached 99.2% (Figure 4.1). These results demonstrate that salt-assisted roasting is an effective strategy for the recovery of spent LFP.[5] However, in the recycling of spent LFP, pyrometallurgy presents three major drawbacks: (1) LFP cannot be directly reduced by carbon to form the Fe olivine structure and requires the addition of molten salts to assist in roasting. (2) The metallurgical temperature typically exceeds 800 °C, and the high-temperature volatility of lithium can lead to the collapse of the $FePO_4$ framework, thereby reducing lithium recovery efficiency. (3) The process is characterized by high energy consumption and low sustainability. In addition, combustion generates volatile and harmful gases, necessitating the use of dedicated flue gas purification systems.

4.1.2 Analysis of Pyrometallurgical Recovery Technology for LFP

The pyrometallurgical recycling process for LFP batteries faces challenges, including low energy efficiency, substantial resource waste, and considerable environmental impact. Firstly, the pyrometallurgical process requires high temperatures (typically 600–850 °C) to decompose LFP cathode materials and extract the key elements—iron (Fe), phosphorus (P), and lithium. However, apart from lithium, the

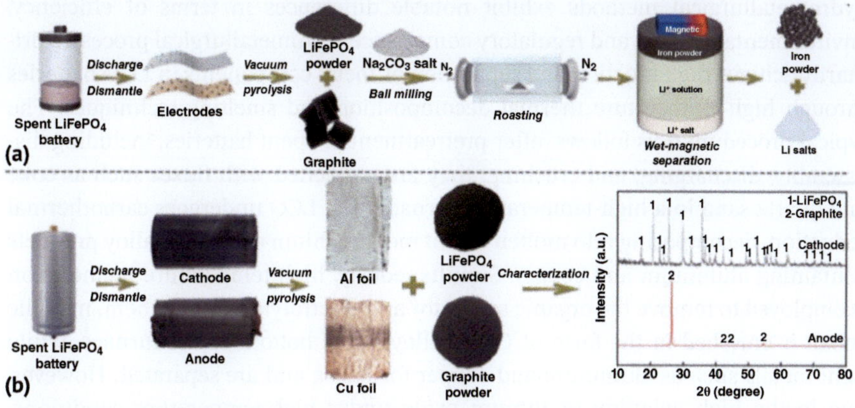

Figure 4.1 (a) Flow diagram for the Na_2CO_3-assisted carbothermic roasting-magnetic separating process and (b) digital photos for treating the spent LFP batteries and XRD patterns of scraped powders from current collectors. *Source:* Adapted from [5].

intrinsic value of other metals in LFP is relatively low, and the market value of the resulting recycling products (such as $FePO_4$ or Li_2CO_3) is also limited, making it insufficient to offset the high cost of high-temperature treatment. The regeneration value of iron-based alloys or phosphates is only about one-quarter that of ternary materials, indicating significantly lower economic viability. In addition, high-temperature treatment can result in irreversible lithium loss. Lithium tends to react with silicon and aluminum oxides in the slag, forming stable compounds (such as Li_2SiO_3), which increases recovery difficulty and contradicts the objective of resource recycling. The environmental burden of pyrometallurgical processes cannot be overlooked. During high-temperature treatment, the organic electrolyte and graphite anode in LFP can release toxic gases such as HF and CO. An additional off-gas purification system is required to control these emissions. However, the treatment efficiency is limited and the operational costs are high. Meanwhile, the high-temperature decomposition process generates a significant amount of CO_2 emissions, which contradicts the goal of carbon neutrality. More critically, the high temperature can destroy the olivine structure of LFP, resulting in poor electrochemical performance of the pyrometallurgically recycled materials. The initial discharge capacity and cycle stability of these materials are significantly reduced, making it difficult to meet the stringent requirements of power batteries. Consequently, they can only be further processed through subsequent hydrometallurgical methods to achieve effective metal extraction.

4.1.3 Application of Pyrometallurgical Recycling in the Lithium Cobalt Oxide ($LiCoO_2$, LCO)

LCO batteries are the dominant power source in the consumer electronics industry. Due to their concentrated resource composition (with cobalt [Co] content exceeding 60%) and the requirement for precise metal extraction, their recycling processes have evolved into distinct technical pathways. Pyrometallurgical and hydrometallurgical methods exhibit notable differences in terms of efficiency, environmental impact, and regulatory compliance. Pyrometallurgical processes primarily achieve the reduction and separation of metal components in LCO batteries through high-temperature thermal decomposition and smelting techniques. The typical procedure is as follows: after pretreatment of spent batteries, including disassembly, discharging, and crushing, they are co-melted with fluxes such as coke and quartz sand in a high-temperature furnace. The LCO undergoes carbothermal reduction, decomposing into molten cobalt metal, lithium oxide, and alloy products containing aluminum and copper. Simultaneously, high-temperature incineration is employed to remove the organic separator and electrolyte. Among them, metallic cobalt is enriched in the form of Co-Fe alloy at the bottom of the furnace, while light metals such as aluminum and copper form slag and are separated. However, due to the high volatility of lithium oxide under high-temperature conditions, approximately 60 to 80% of the lithium is dispersed in the form of lithium oxide in the furnace slag and requires purification through subsequent hydrometallurgical leaching. Compared with hydrometallurgy, the core advantages of pyrometallurgy lie in its large processing capacity, with a single furnace capable of handling 10 tons per day; its short process flow, which eliminates the need for complex leaching and extraction steps; and its strong raw material compatibility, enabling direct processing of batteries with mixed chemical compositions or severe oxidation. However, it faces challenges such as high thermal energy consumption, the potential release of harmful elements like fluorine and phosphorus under high-temperature conditions and low lithium recovery efficiency. In terms of environmental pressures, modern pyrometallurgical processes have reduced the concentration of pollutant gases through pretreatment for fluorine removal, optimization of smelting conditions, and multistage exhaust gas purification. When integrated with waste heat power generation systems utilizing flue gas, these processes effectively lower energy consumption costs. Furthermore, by adjusting the composition of the slag system, the total lithium recovery rate can be enhanced to over 75%. GEM employs a coupled pyrometallurgical-hydrometallurgical process to treat spent LCO batteries. After pretreatment, the materials are smelted in an electric arc furnace at 1,600 °C with coke and quartz sand to produce Co-Fe alloy and lithium-rich slag. Aluminum and copper are separated and recovered as valuable by-products. The lithium-containing slag is water-quenched and subsequently leached with sulfuric acid, followed by lithium precipitation. Flue gas is rapidly cooled and scrubbed with an alkaline solution to minimize fluoride emissions. The residual heat is recovered to supply approximately 30% of the total energy demand. The single-line daily processing capacity reaches 60 tons, with cobalt recovery rate exceeding 95%. This process is particularly suitable for handling mixed and severely damaged batteries. Although lithium recovery depends on hydrometallurgical methods, large-scale production has reduced the overall cost by 25% compared to purely hydrometallurgical approaches. Currently, researchers have also carried out related investigations. Maroufi et al. developed a carbothermal reduction–cold water leaching process to recover lithium and cobalt from spent LCO cathodes. After carbothermal reduction at 700 °C under argon atmosphere for 60 minutes, LCO decomposed into Li_2CO_3, CoO, and metallic Co. Subsequently, Li_2CO_3

was selectively leached using cold water at 0 °C, resulting in a lithium recovery rate of 36%. The residual CoO was further reduced at 700 °C to produce metallic cobalt with purity exceeding 99%. This process employs low-temperature carbon thermal fixation to convert lithium into lithium carbonate, thereby preventing volatilization losses at elevated temperatures. It integrates selective cold water leaching to enable the green separation of lithium and cobalt. The entire process is free of strong acids and features a simple, straightforward operation.[6] Xiao et al. proposed a vacuum pyrolysis method for recovering lithium carbonate from spent lithium manganate ($LiMn_2O_4$) cathodes. The electrode material was pyrolyzed at 973 K under vacuum (<1 kPa) for 30 minutes, during which lithium was transformed into Li_2CO_3. Subsequent water leaching achieved a lithium recovery rate of 81.9%, with a purity reaching up to 99.7%. This method is also applicable to LCO and $LiNiCoMnO_2$ batteries, achieving lithium recovery rates of 82.7 and 66.3%, respectively. Economic assessment indicates that processing 10 tons of spent batteries generates a daily profit of approximately $2,108, demonstrating that the process is both environmentally sustainable and economically feasible.[7] Yang et al. proposed a novel process integrating reductive thermal treatment and acid leaching without the addition of a reductant to recover valuable metals from spent LCO batteries. The recovery rates of lithium and cobalt approached 100%. The proposed approach can not only make full utilization of waste anode graphite, but also benefit from leaching valuable metals in the absence of reductant, which significantly improves the economy and recovery performance of recycling spent LIBs (Figure 4.2b).[8]

4.1.4 Analysis of Pyrometallurgical Recovery Technology for LCO

The limitations of pyrometallurgy in the recycling of LCO batteries:

Although the pyrometallurgical recycling process offers certain advantages in cobalt metal extraction from LCO batteries, it has several technical limitations, primarily in terms of energy consumption efficiency, resource utilization comprehensiveness, and environmental impact. Firstly, the core of pyrometallurgical processes

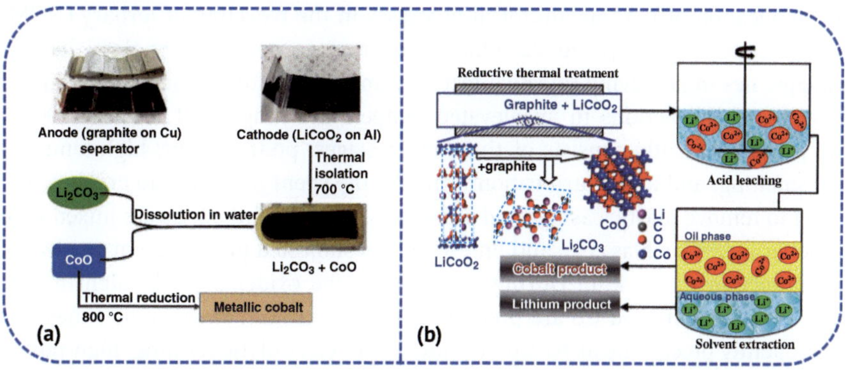

Figure 4.2 (a) Recovery of lithium and cobalt from waste LIBs through a selective isolation-suspension approach. *Source:* [6] / with permission of Elsevier. (b) Recovering valuable metals from spent LIB via a combination of reduction thermal treatment and facile acid leaching. *Source:* [8] / with permission of American Chemical Society.

involves high-temperature roasting (typically above 1,300 °C), but lithium in LCO has a relatively low boiling point. Under such high temperatures, lithium may volatilize in the form of lithium carbonate or lithium oxide and enter the flue dust, leading to a low overall lithium recovery efficiency. This limitation contradicts the current high demand for lithium resources. Secondly, although the pyrometallurgical process can effectively concentrate cobalt into metal alloys, the aluminum current collector reacts vigorously with oxygen during smelting, forming aluminum oxide slag. This reaction not only erodes the furnace lining and increases equipment maintenance costs but also encapsulates some cobalt particles, necessitating subsequent extraction via secondary acid leaching, which indirectly prolongs the process flow. Another potential risk arises from the treatment of organic components—the electrolyte and separator in batteries can generate highly toxic fluorine-containing gases (e.g., HF) and dioxins during incineration, potentially leading to secondary environmental pollution. Meanwhile, the pyrometallurgical process imposes extremely stringent requirements on the homogeneity of raw materials. However, in practice, recovered LCO batteries are often mixed with other battery types or different generations of products (e.g., retired batteries containing lithium manganate, ternary lithium, or LFP). These impurities may form complex intermetallic compounds with cobalt during roasting, leading to reduced cobalt purity and diminished crystal activity, which often fail to meet the standards required for high-value recycled materials. Furthermore, the pyrometallurgical process exhibits reduced efficiency when processing fine LCO powder. Some of the powder that remains unmelted may be entrained by the gas flow and transported into the dust removal system as fly ash, thereby indirectly increasing the cobalt loss rate. Overall, although pyrometallurgy simplifies the metal separation process, the elemental losses caused by high temperatures, along with associated environmental risks and performance deficiencies in recycled materials, have increasingly led to its replacement in the field of high-efficiency recycling by greener and more sustainable technologies.

4.1.5 Application of Pyrometallurgical Recovery in Ternary Batteries

The application of pyrometallurgical processes in the recycling of ternary (NCM) LIBs is relatively widespread. As a high-temperature smelting technology, its core principle lies in achieving selective recovery and separation of metal components through redox reactions in an elevated temperature range of 1,300 to 1,600 °C. The process typically consists of three main stages: pretreatment, high-temperature smelting, and alloy purification. Initially, the spent batteries are crushed and sorted to remove plastic casings and separators. Subsequently, they are mixed with reducing agents such as coke and limestone and subjected to roasting in an electric arc furnace or rotary kiln. During this process, the oxides of high-valent metals such as nickel (Ni) and Co are preferentially reduced to their metallic states and subsequently precipitate as Ni-Co-Fe alloys. After electrolytic refining, high-purity, battery-grade nickel sulfate and cobalt sulfate products can be obtained. However, manganese (Mn) and Li exhibit distinct behavioral differences due to variations in their thermodynamic properties. Approximately 80% of manganese reacts with iron, aluminum, and other metals to form a glassy furnace slag, while lithium volatilizes into gaseous Li_2O at high temperatures or accumulates on the surface of the slag. In traditional processes, less than 15% of lithium can be effectively recovered.

Despite its relatively low lithium recovery efficiency, pyrometallurgy still offers significant industrial advantages: each production line is capable of processing 50–80 tons per day, making it especially suitable for the direct treatment of black powder with complex compositions that are difficult to preprocess. Moreover, the process does not require precise material sorting and exhibits a high tolerance for impurities in raw materials. Current technological innovations primarily aim to improve metal recycling efficiency and optimize energy consumption. For example, the microwave-assisted dynamic smelting system developed by the US-based company Recupyl, which operates at temperatures between 1,800 and 2,000 °C, has enhanced lithium vapor recovery efficiency to 58% through the selective heating mechanism. The molten salt electrochemical refining process developed by Germany's Fraunhofer Institute allows for the direct electrodeposition of Co-Ni alloys in the molten NaCl-KCl electrolyte, achieving a 37% reduction in energy consumption compared to traditional pyrometallurgical methods. Consistent with the large-scale industrial application of pyrometallurgy, laboratory-level research on recycling based on pyrometallurgical principles has led to the development of various innovative improvement models. Zhang et al. successfully recovered lithium and manganese from spent NCM batteries by the combination of controlled carbon thermal reduction, selective leaching, and precipitation techniques. The spent $LiNi_xCo_yMn_{1-x-y}O_2$ was first roasted to obtain Ni-Co alloy, MnO, Li_2O, and Li_2CO_3 at 750 °C for 3 hours by graphite recycled from the spent anode. Subsequently, H_3PO_4 was applied to selectively leach 99.1% of Li and 96.3% of Mn at 40 °C for 10 minutes, while merely 4.5% of Co and 1.2% of Ni were dissolved. Finally, precipitation of the leaching solution converts lithium and manganese into final products such as Li_3PO_4 and $Mn_3(PO_4)_2{\cdot}3H_2O$.[9] Zhang et al. aimed to explore the thermal reduction treatment of spent $LiNi_xCo_yMn_zO_2$ cathode materials from LIBs by utilizing the spent anode graphite as a reductant, followed by the leaching of valuable metals without the use of additional reductive agents. At 600 °C and with a cathode-to-anode material mass ratio of 6:1, the mixed electrode materials can be converted primarily into the desired phases of CoO, NiO, MnO, and Li_2CO_3. Being different from obtaining Co and Ni metallic phases in reduction roasting, producing CoO and NiO benefits a lower energy consumption, no H_2 emission in the leaching process, and more facile conditions for complete leaching.[10] Peng et al. introduced steam into the reduction roasting process. The anode graphite was efficiently gasified by the steam, producing hydrogen and carbon monoxide (CO), thereby significantly accelerating the reduction of CoO and NiO while simultaneously reducing the carbon content in the calcination products. Under conditions of H_2O/C molar ratio of 5:1 and a reduction calcination temperature of 850 °C, the magnetic separation efficiencies of cobalt and nickel in the subsequent process reached 90 and 93%, respectively, and the final recovered products were high-purity Co, Ni, and Li_2CO_3.[11] Chen et al. a new process for selective extraction of lithium from spent LIBs by roasting a mixture of $LiNi_{0.8}Co_{0.1}Mn_{0.1}O_2$ and $NaHSO_4{\cdot}H_2O$ followed by water leaching. During the calcination process, $NaHSO_4$ decomposes to generate SO_3, which promotes the conversion of lithium into the water-soluble compound $LiNaSO_4$, whereas NCM remains in the form of oxide or sulfate complex salts. Under optimal conditions, the lithium leaching rate reaches 99.3%, while the leaching rates of NCM remains below 3.9%. The overall process does not require the use of strong acids and is characterized by lower energy consumption.[12]

4.1.6 Analysis of Pyrometallurgical Recovery Technology for NCM

Although the pyrometallurgical recycling process enables the efficient recovery of high-value metal elements such as cobalt and nickel from NCM batteries, it is accompanied by significant drawbacks, including high energy consumption, low lithium recovery efficiency, and considerable environmental costs.

The limitations of pyrometallurgical processes in the recycling of NCM batteries are closely linked to their fundamental characteristic: high-temperature smelting and roasting. Firstly, when the cathode material of NCM batteries is subjected to high-temperature treatment in an electric arc furnace or rotary kiln, fluorine and phosphorus elements released from the decomposition of organic materials such as aluminum foil, copper foil, and separators can readily form complex eutectic compounds with metals, leading to the production of high-value metals such as nickel and cobalt in alloy form. The separation of this mixed metal requires the implementation of additional processes such as acid leaching or electrolysis, thereby increasing the complexity of the overall process. Secondly, in NCM batteries, lithium elements are highly susceptible to reacting with impurities in the furnace at temperatures exceeding 1,500 °C, forming low-activity compounds such as lithium silicate, which results in a lithium recovery rate typically below 50%. What makes the situation more complex is that the pyrometallurgical process is highly sensitive to variations in the chemical composition ratios of battery precursor materials. Different NCM models (such as NCM111, NCM523, and NCM811) or the presence of mixed LFP battery fragments can alter the component distribution within the molten pool, leading to fluctuations in the composition of the final smelting product and significantly compromising the purity and electrochemical consistency of the recycled materials. In addition, the decomposition of NCM battery electrolytes generates highly toxic gases such as HF and P_2O_5, necessitating the installation of multistage rapid cooling, alkali washing, and activated carbon adsorption systems. Furthermore, residual fluorides tend to form coke deposits on the inner walls of equipment, significantly increasing both the frequency of system maintenance and the cost of tail gas treatment. Compared to the hydrometallurgical process, which enables highly efficient recovery of elemental metals, the pyrometallurgical process—based on the "melt first, then purify" strategy—has long been challenged by excessively high costs when handling the high heterogeneity of NCM battery components.

4.1.7 Application of Pyrometallurgical Recycling in Solid-state Batteries

The pyrometallurgical recycling process for solid-state batteries differs significantly from that of traditional LIBs, with the core contradiction arising from the chemical properties of the solid electrolyte and the unique structure of the battery itself. Traditional LIBs are characterized by the presence of the liquid organic electrolyte and polyolefin separator. During pyrometallurgical recycling, organic substances are primarily decomposed at high temperatures (1,200–1,500 °C), while metal alloys (such as cobalt, nickel, and copper) are concentrated. However, solid-state batteries utilize solid electrolytes (such as oxide $Li_7La_3Zr_2O_{12}$ [LLZO], sulfide $Li_{10}GeP_2S_{12}$, or polymer PEO-LiTFSI), and their high-temperature pyrolysis behavior, forms of metal existence, and pollutant generation mechanisms all undergo fundamental changes. Solid-state electrolytes exhibit higher thermal stability, with decomposition temperatures 200–400 °C above those of liquid electrolytes, greater compositional complexity, containing rare earth elements, sulfur, germanium, and other components, and stronger interfacial bonding with electrode materials. These characteristics necessitate modifications to temperature profiles, off-gas treatment systems, and metal separation pathways in pyrometallurgical recycling processes. In addition, solid-state batteries lack liquid electrolytes and separators, thereby avoiding the concentrated release of highly toxic substances such as HF and dioxins during traditional pyrometallurgical recycling. However, the decomposition of sulfide electrolytes may generate new pollutants such as H_2S and SO_2. Therefore, hydrometallurgical separation and recycling methods are currently predominantly employed for polymer-based and sulfide-based solid-state batteries. Meanwhile, the development of targeted exhaust gas purification technologies is essential to enable effective pyrometallurgical recovery. Currently, pyrometallurgical recovery of solid-state batteries is primarily focused on oxide-based solid-state batteries.

Oxide-based solid electrolytes demonstrate superior thermal stability. Upon contact with electrodes, the initial decomposition temperatures of all solid electrolytes decrease, although the reaction mechanism between the solid electrolytes and the electrodes remains unclear. For sulfide- and polymer-based solid-state batteries, the solid-state electrolyte is either combusted or converted into slag during high-temperature calcination. This not only prevents the recycling of the solid-state electrolyte but also results in the generation of toxic gases. The oxide cathode in the aforementioned battery system is converted into the alloy or slag phase, a process that closely resembles the recycling mechanism of spent LIBs. For oxide-based solid-state batteries, the solid electrolyte shares a similar chemical composition with the cathode material, and the metals in the battery (such as lithium, gallium, zirconium, tantalum, manganese, and nickel) are converted into alloys or slags during high-temperature calcination. Subsequently, the calcined products undergo further separation and recovery through a sophisticated hydrometallurgical process.[13] However, the diversity of metals in oxide-based solid-state batteries presents substantial challenges for metal separation and recycling processes. This battery recycling technology offers advantages such as high industrial maturity. However, complex operational procedures, low battery recycling rate, and the emission of harmful gases and wastewater collectively contribute to limited industrial potential for solid-state battery recycling. In summary, the technology of extracting valuable metals from spent batteries by disrupting the crystal structure of the solid electrolyte cathode is not applicable to solid-state batteries.

4.1.8 Analysis of Pyrometallurgical Recovery for Solid-state Technology

Pyrometallurgical recycling processes enrich valuable metals from solid-state batteries through high-temperature smelting and reduction reactions, but the main technical bottlenecks lie in the sharp increase in energy consumption, limited metal separation efficiency, and the challenges associated with treating emerging pollutants.

The high thermal stability and compositional complexity of solid electrolytes have substantially modified the reaction pathways of conventional pyrometallurgical processes. For instance, the melting point of oxide-based solid electrolytes (such as LLZO) ranges from 1,650 to 1,700 °C, significantly exceeding the decomposition temperature of LCO in conventional LIBs. This necessitates an increase in smelting temperature above 1,600 °C. During this process, LLZO undergoes a solid-state reaction with the cathode material (e.g., $LiCoO_2$), forming high-melting-point compounds (e.g., $LaCoO_3$), which immobilize cobalt and lithium within the refractory slag phase. To enhance the fluidity of the slag phase, additional fluxes must be introduced; however, this also brings calcium and silicon impurities, further reducing the metal grade. In addition, rare earth elements (e.g., lanthanum and zirconium) and transition metals (e.g., cobalt and nickel) in solid electrolytes form multicomponent alloy phases at high temperatures. Conventional pyrometallurgical methods, such as density separation or magnetic separation, are ineffective for efficiently separating these elements. As a result, hydrometallurgical acid leaching must be employed for secondary purification, which further increases process complexity. The introduction of sulfide-based solid electrolytes (e.g., Li_3PS_4) introduces new chemical risks. During high-temperature smelting, the concentration of gaseous pollutants, such as H_2S and SO_2, generated from the decomposition of sulfides increases, necessitating the implementation of the multistage purification system, which in turn raises the capital investment cost of the equipment. Meanwhile, sulfur reacts with metals (e.g., cobalt and nickel) to form sulfides (CoS and NiS), which have melting points lower than those of the corresponding metallic elements. This results in poor phase separation between the molten metal and sulfur slag, leading to a lower metal recovery rate compared to conventional pyrometallurgical methods. Polymer-based electrolytes (e.g., PEO-LiTFSI) undergo carbonization at high temperatures, forming fluorinated carbon (CF_x), which not only results in the loss of lithium encapsulation but also necessitates the use of specialized exhaust gas treatment modules due to the toxicity and greenhouse potential of their pyrolysis products (e.g., HF and CF_4), thereby increasing operational and maintenance costs.

Additionally, the dense structure of solid-state batteries further complicates the pretreatment process. The conventional pyrometallurgical process, which relies on high-temperature crushing methods (e.g., electric arc furnace smelting), is inefficient for dismantling the integrated design of solid-state batteries (e.g., all-solid-state laminated structures). The physical integration of electrodes and electrolytes results in non-uniform metal enrichment, necessitating the introduction of pretreatment technologies such as high-pressure roller grinding or liquid nitrogen-based deep cryogenic crushing, thereby increasing energy consumption compared to conventional processes. However, the market demand for rare earths (such as lanthanum [La] and zirconium [Zr]) in the smelting products is limited, and the added value of recycled metals is low, making it difficult to cover the separation costs, with economic bottlenecks being prominent. Ultimately, the pyrometallurgical process encounters three major challenges in the recycling of solid-state batteries: poor high-temperature adaptability, low metal recovery rate, and high pollution control costs. Therefore, it is imperative to develop low-temperature fluxes, targeted separation techniques, and waste gas resource utilization technologies to overcome the barriers to large-scale application.

4.1.9 Demand for Pyrometallurgical Recycling Technology Targeting Cathode Materials from Spent LIBs

To overcome the current limitations of pyrometallurgical recycling technology in terms of environmental sustainability and high-value recovery, it is imperative to propose a set of systematic technical requirements based on its key shortcomings, thereby guiding future technological optimization and process route design. Firstly, it is essential to substantially enhance lithium recovery efficiency. Lithium, as an indispensable key metal in battery applications, experiences particularly significant loss during the pyrometallurgical process. It is essential to develop process technologies that can suppress lithium volatilization and promote its enrichment in the metallic phase under high-temperature conditions or enable efficient enrichment and one-step recovery of lithium through strategies such as selective reduction and flux control, thereby reducing reliance on hydrometallurgical methods. Secondly, it is imperative to reduce energy consumption and carbon emissions during the high-temperature smelting process. The traditional pyrometallurgical process relies on carbon-based reducing agents and high-temperature operations, which result in substantial energy consumption and significant carbon emissions, thereby misaligning with the goal of carbon neutrality. Therefore, pyrometallurgy should transition toward a low-temperature and environmentally sustainable reduction pathway, encompassing the adoption of electric heating to replace traditional fuels, the development of hydrogen-based and other carbon-free reduction systems, and the optimization of furnace structures to enhance thermal energy utilization efficiency, with the objective of significantly reducing both unit energy consumption and carbon footprint. Thirdly, it is essential to systematically improve the elemental recovery coverage and resource utilization efficiency of pyrometallurgy, while enhancing the recovery capacity of metals with low reducibility or low added value, such as manganese and aluminum. Therefore, it is essential to develop auxiliary extraction processes or slag phase regulation technologies to enable efficient recovery with reduced energy consumption. For nonmetallic resources such as anode graphite, electrolyte components, and organic binders, pyrometallurgical processes still lack systematic treatment pathways. It is essential to explore technologies for collecting and utilizing high-temperature pyrolysis products in order to achieve a fully closed-loop recycling process. Fourth, it is crucial to strengthen environmental safety control throughout the pyrometallurgical process. The battery contains organic components that release toxic gases, such as HF, CO, and VOCs, under high-temperature conditions, necessitating the use of additional exhaust gas treatment systems. Therefore, it is essential to establish an efficient and cost-effective multistage exhaust gas purification system and implement real-time coordinated control with the main process, ensuring compliance with pollutant emission standards and guaranteeing operational safety. Finally, to address the increasing complexity of spent battery component structures and the high costs associated with classification and dismantling, pyrometallurgical recycling should demonstrate greater adaptability to varying components and enhanced process synergy. In future development, the functional role of pyrometallurgy as a pretreatment or enrichment unit should be emphasized. By integrating it with hydrometallurgy, mechanical disassembly, bioleaching, and other processes, a flexible, efficient, and low-emission integrated recycling system should be established to enhance the overall recycling efficiency and economic viability. In summary, the future development of pyrometallurgical recycling technology should focus on the four core dimensions of "low energy consumption, high recovery efficiency, low emissions, and process synergy," break through current technical bottlenecks, and drive the lithium battery recycling industry toward the objectives of green sustainability, high value-added recovery, and large-scale implementation.

4.2 Hydrometallurgical Extraction Technology

4.2.1 Application of Hydrometallurgical Recycling in the LFP

Hydrometallurgy is one of the mainstream technologies for LFP battery recycling. Through chemical dissolution and selective purification, it enables the efficient recovery of lithium, iron, and phosphorus from LFP batteries. In the pretreatment stage, disassembled and crushed LFP batteries (electrodes or black powder) undergo physical sorting to remove aluminum foil, copper foil, and separator residues. Subsequently, acidic reagents are employed as leaching agents, while H_2O_2 and other reducing agents are introduced as auxiliary reagents to enhance lithium dissolution. During this process, LFP is converted into Li^+, Fe^{2+}/Fe^{3+}, and PO_4^{3-} ions in the liquid phase, with the lithium leaching rate exceeding 95%, which demonstrates significantly superior performance compared to the pyrometallurgical process. Subsequently, the elements are separated through methods such as pH gradient regulation, precipitation, or solvent extraction. If high-purity lithium salts are required, they can be further purified through ion exchange or electrolytic refining processes. Its advantage lies in the high directional recovery efficiency of lithium and LFP ($Li > 90\%$, $Fe/P > 85\%$), and the regenerated materials (Li_2CO_3 and $FePO_4$) can be directly utilized in the production of new batteries, thereby enabling a closed-loop recycling system. The current main approach is to selectively extract lithium first, followed by the recycling of iron and phosphorus slag.

4.2.1.1 Selective Lithium Extraction

The acid consumption and impurity interference issues are particularly pronounced during the leaching stage. Residual aluminum foil in LFP cathode materials dissolves

Table 4.1 The primary selective lithium extraction methods.

Method	Reagent	Leaching efficiency	References
Oxidative leaching	CH_3COOH, H_2O_2	Li: 95.05%, Fe: 5.92%	14
Oxidative leaching	H_2SO_4, H_2O_2	Li: 96.85%, Fe: 0.027%, P: 1.95%	15
Oxidative leaching	$Na_2S_2O_8$	Li: 99%, Fe: 0.048%, P: 0.387%	16
Oxidative leaching	Air, H_2SO_4	Li: 99.3%, Fe: 0.02%	17
Electrochemical approach	NaCl	Li: 98%, Fe: 4%	18
Electrochemical approach	Li_2SO_4, LiOH	Li: 96.74%, Fe: 4.13%, P: 3.98%	19
High-temperature phase conversion	H_2SO_4, H_2O_2	Li: 98.46%, Fe: 0.01%, P: 26.59%	20
Mechanochemical approach	$H_2C_2O_4$	Li: 99%, Fe: 6%	21

to release Al^{3+} under acidic conditions. The $AlPO_4$ gel formed by Al^{3+} and PO_4^{3-} can encapsulate the active materials, thereby inhibiting lithium release. Consequently, the introduction of additional chelating agents (e.g., citric acid) or buffer systems is necessary to mitigate this issue. Another recycling challenge stems from the strong chemical bonds between lithium and iron or phosphorus. The conventional hydrometallurgical leaching process requires high-concentration acid and elevated temperatures to release lithium; however, the simultaneous dissolution of impurities such as Fe^{3+} and PO_4^{3-} significantly interferes with subsequent purification steps. Selective lithium extraction technology, which enables efficient lithium recovery by precisely controlling the speciation and reaction pathways of lithium while minimizing impurity dissolution, has emerged as a research hotspot in the field of LFP battery recycling. The primary objective is to selectively extract the higher-value lithium from spent LFP through methods such as oxidation and mechanical activation transformation, thereby ensuring a high lithium recovery efficiency.

4.2.1.1.1 Oxidative Leaching

The oxidative leaching refers to a process wherein an oxidant is introduced during hydrometallurgical leaching to convert Fe^{2+} within LFP crystals into Fe^{3+}. This transformation facilitates the selective release and leaching of lithium embedded in the crystal lattice from LFP, thereby enabling efficient separation of lithium and iron. Among these, H_2O_2 is the most commonly used and widely adopted oxidant. Yang et al. employed H_2O_2 as the oxidant and compared the leaching selectivity of oxalic acid, citric acid, acetic acid, and hydrochloric acid for lithium recovery. Acetic acid was ultimately selected as the leaching agent. Under the optimized conditions (0.8 mol/L CH_3COOH, 6% H_2O_2, liquid-to-solid ratio of 120 g/L, 50 °C, 30 minutes), approximately 95.05% of lithium was leached. Subsequently, Na_2CO_3 was added to precipitate lithium, resulting in the obtainment of battery-grade Li_2CO_3. The residual iron and phosphorus remain in the form of $FePO_4$ within the leaching residue.

Controlling the oxidation state of the solution is crucial for selective lithium extraction. It is essential to oxidize Fe^{2+} to Fe^{3+} while preventing the dissolution of iron into the solution.[14] Li et al. employed low-concentration sulfuric acid (H_2SO_4) for leaching with H_2O_2 as the oxidant. Under optimized conditions (0.3 mol/L sulfuric acid, $n(H_2SO_4):n(Li) = 0.57$, 60 °C, and 2 hours), the lithium leaching efficiency reached 96.85%. Over 99% of iron and 98% of phosphorus were retained in the residue as $FePO_4$. Following filtration, Na_3PO_4 was added as a precipitating agent to recover lithium.[15] Kumar et al. proposed a lemon juice–H_2O_2 synergistic process that, under optimal conditions, enables selective leaching of 94.8% of lithium from spent LFP, with less than 4% Fe loss. The leaching solution was utilized for lithium precipitation to obtain Li_2CO_3 with a purity of 99.2%. The leaching residue was calcined and recycled for LFP cathode regeneration, which exhibited a capacity of 155 mAh/g at 0.1C and a retention rate of 98.3% after 100 cycles at 1C.[22]

In addition to employing H_2O_2 as an oxidant, some researchers have also explored the use of alternative oxidants for oxidative leaching. Zhang et al. oxidized Fe^{2+} to Fe^{3+} by employing $Na_2S_2O_8$ as the oxidant, without the addition of acid or alkali. The effects of liquid-to-solid ratio, oxidant dosage, temperature, and reaction time on the selective leaching efficiency of lithium were investigated. Under conditions of 1.05 times the theoretical $Na_2S_2O_8$ dosage, a liquid-to-solid ratio of 300 g/L, a reaction time of 20 minutes and ambient temperature, the lithium leaching efficiency reached up to 99%, while the leaching efficiencies of Al, Fe, and P were only 0.584, 0.048, and 0.387%, respectively. The purity of the recovered Li_2CO_3 exceeded 99%, making it suitable for direct use in the synthesis of battery materials.[16] Shentu et al. proposed an ammonium persulfate oxidation process for the efficient and selective recovery of lithium from spent LFP. Under optimal conditions, the lithium leaching efficiency reaches 99%, with Fe remaining in the system as $FePO_4$ and the olivine structure remaining intact. The leaching solution is subjected to electrolysis for LiOH production, and the mother liquor can be concentrated to regenerate ammonium persulfate, thereby establishing a closed-loop process. This process introduces no additional impurities and demonstrates significantly higher energy efficiency compared to traditional acid leaching.[23] In addition, to reduce oxidant costs, some researchers have further proposed an air oxidation-based selective lithium extraction process, with its mechanism illustrated in Figure 4.2a. Selective lithium extraction can be achieved by utilizing green and cost-free air as the oxidant in combination with sulfuric acid for pH control, resulting in a lithium leaching efficiency of 99.3%.[17] Yan et al. proposed an HCl-NaClO synergistic leaching process that selectively extracted 99% of lithium from spent LFP (with less than 2% Fe dissolution) by precisely controlling the dosages of HCl and NaClO as well as the pH conditions. Subsequently, lithium in the leachate was recovered as Li_2CO_3 through solvent extraction and precipitation, achieving an overall recovery efficiency of 97%. This process reduces acid and alkali consumption by 50% compared to traditional methods, eliminates the need for highly toxic oxidants, and preserves the olivine structure of LFP, offering a promising approach for the sustainable recycling of low-value LFP batteries.[24] Wu et al. proposed a process for recovering waste LFP materials via oxidative pressure leaching, employing O_2 as the oxidant and stoichiometric

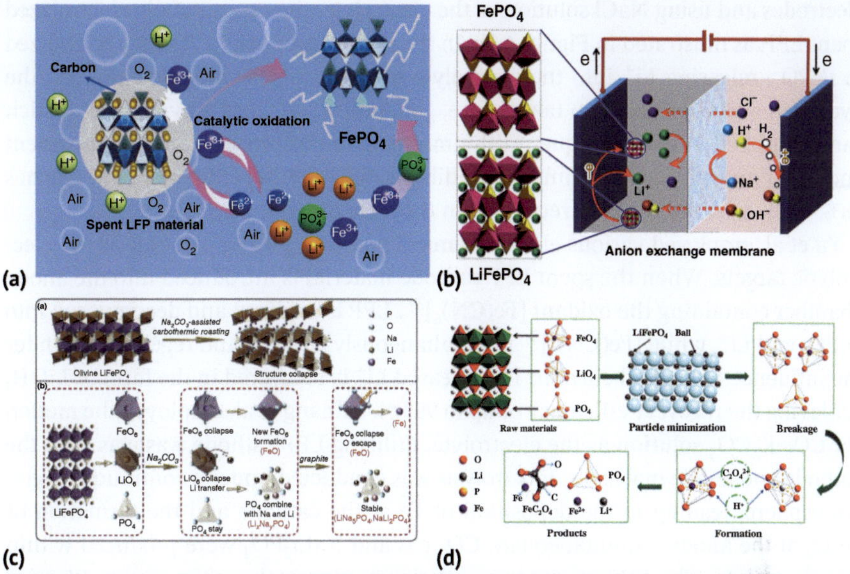

Figure 4.3 (a) Schematic diagram of the leaching mechanism of the air pressure leaching. *Source:* [17] / with permission of Elsevier. (b) Schematic diagram of the slurry electrolysis. *Source:* [18] / with permission of Elsevier. (c) Schematic illustration of the conversion mechanism for the Na_2CO_3-assisted carbothermic roasting of LFP. *Source:* Adapted from [5]. (d) Possible products and mechanisms in the mechanochemical process. *Source:* [21] / with permission of American Chemical Society.

sulfuric acid as the leaching agent (Figure 4.3a). More than 97% of lithium was transferred into the solution, whereas over 99% of iron remained in the leaching residue, enabling a one-step separation of Li and Fe. Subsequently, by adjusting the pH of the leachate, phosphate and lithium were precipitated, leading to the recovery of over 95% of lithium in the form of Li_3PO_4.[25] Zhou et al. proposed a phosphoric acid-based synergistic leaching process. By optimizing the H_3PO_4 concentration, reaction temperature and reaction time, Ni (99.1%), Co (98.9%), Li (99.6%), and Mn (97.3%) were simultaneously recovered in a single step from mixed spent batteries, while Fe was selectively precipitated as $FePO_4{\cdot}2H_2O$ with a recovery efficiency of 98%. This process leverages the spontaneous redox reaction between Fe^{2+} and NCM cathode materials, eliminating the need for external oxidants and pH adjustment, and reducing acid consumption by 60%.[26]

4.2.1.1.2 Electrochemical Approach

The electrochemical method is powered by electrical energy and enables lithium extraction through chemical reactions that occur at the electrodes. The current acts as an equivalent to the “oxidizing agent” in the leaching process, driving lithium ions to leach from the original structure into the electrolyte. Li et al. integrated the charging and discharging mechanism of LFP batteries with slurry electrolysis technology. By employing an anion exchange membrane to separate the positive and negative electrodes and using NaCl solution as the electrolyte, they successfully electrolyzed spent LFP, as illustrated in Figure 4.3b. In the anode chamber, LFP slurry is oxidized to $FePO_4$, releasing Li^+ into the electrolyte solution. In the cathode chamber, the hydrogen evolution reaction takes place, generating NaOH as a by-product, which can be applied in the subsequent lithium precipitation process for pH adjustment and iron removal. Under optimized conditions, the leaching efficiency of Li reaches 98%, while over 96% of Fe is recovered in the form of $FePO_4$/C.[18]

Yu et al. employed various electrochemical processes, using oxidants as the electrolytic targets. When the spent LFP cathode material is introduced into the anode chamber containing the oxidant $[Fe(CN)_6]^{3-}$, LFP is oxidized and decomposed into $FePO_4$ and Li^+, while $[Fe(CN)_6]^{3-}$ is simultaneously reduced and regenerated under the influence of applied current. The released Li^+ is recovered in the form of LiOH, achieving the recovery efficiency of up to 99.8%.[27] Zhang et al. employed the molten Na_2CO_3-K_2CO_3 solution as the electrolyte, utilizing LFP cathode waste as both the cathode and anode materials. Electrolysis was conducted under a continuous argon atmosphere, leading to the deposition of Fe at the cathode and the formation of Fe_3O_4 at the anode. Simultaneously, CO_2 gas and $NaLi_2PO_4$ were produced within the electrolyte. The lithium recovery efficiency ultimately achieved was 95.2%.[28] The electrochemical selective lithium extraction method exhibits low reagent consumption and environmental friendliness. However, it requires the waste material to be recoated onto the electrode, a process that is both labor-intensive and time-consuming. Additionally, current research on this method indicates that power consumption remains relatively high, while equipment scalability is limited. Consequently, the method lacks competitiveness in terms of total cost and potential for industrial application.

4.2.1.1.3 High-temperature Phase Conversion

High-temperature phase transformation involves converting lithium in LFP into a more soluble phase under elevated temperatures, enabling selective lithium extraction by exploiting differences in leaching behavior between phases. Tao et al. first oxidized spent LFP cathode material in air at 600 °C, converting it into $Li_3Fe_2(PO_4)_3$ and Fe_2O_3. Subsequently, H_2SO_4 was employed to selectively leach Li^+ and a portion of PO_4^{3-} into the solution, achieving leaching efficiencies of 98.48% for Li, 0.01% for Fe, and 26.59% for P. Finally, lithium was recovered in the form of Li_2CO_3. An appropriate amount of H_3PO_4 is added to the leaching residue, followed by the heating treatment, after which iron is recovered in the form of $FePO_4$.[20] Zhang et al. proposed a method for selective lithium extraction from LFP by leveraging the reducing properties of anode graphite, with the mechanism illustrated in Figure 4.3c. LFP powder and graphite were mixed, and sodium carbonate was added. Subsequently, the mixture was calcined in a tube furnace under a nitrogen atmosphere. As a result, lithium was selectively converted into soluble $LiNa_5(PO_4)_2$ and insoluble $NaLi_2PO_4$, while iron remained in the form of elemental iron. Afterwards, magnetic separation is applied to separate iron from lithium, followed by water leaching to extract the soluble lithium fraction. Insoluble $NaLi_2PO_4$ is subsequently converted into Li_3PO_4 precipitate. The overall lithium recovery rate reaches as high as 99.2%.[5]

4.2.1.1.4 Mechanochemical Approach

Mechanical activation refers to the physical and chemical changes, such as phase transformation, structural defects, strain, and amorphization induced in LFP cathode materials by mechanical forces. These changes enhance the hydrophilicity and interfacial reactivity of the materials, reduce the activation energy of reactants, enable selective lithium leaching from LFP under relatively mild conditions and facilitate reactions that are theoretically difficult to occur.

Fan et al. employed oxalic acid as the grinding agent, with a mass ratio of oxalic acid to spent LFP cathode material of 1:1. After ball milling for 2 hours, water leaching was performed. The experimental results showed that more than 99% of lithium and 6% of iron were leached into the solution. The leaching residue was primarily composed of $FeC_2O_4{\cdot}2H_2O$. Compared with direct hydrometallurgical leaching using oxalic acid as the leaching agent, the mechanical activation process significantly reduced the amount of acid required.[21] The mechanism is illustrated in Figure 4.3d.[29] Liu et al. leveraged the similar outer electron configuration and coordination environment of Na and Li, employing low-cost NaCl as the grinding agent. By applying mechanical force, they induced a solid-state reaction that enabled isomorphic substitution of Li by Na in LFP. After mechanical activation, leaching with Na_2CO_3 solution results in Li_2CO_3 precipitation and enables simultaneous recovery of NaCl.[30]

Compared with other selective lithium extraction methods, the mechanical activation method requires lower amounts of acid and alkali; however, it significantly increases energy consumption during the recovery process. In addition, the combination of mechanical activation and leaching complicates the recycling process.

Lithium is the most valuable component in LFP cathode materials, and efficient lithium recovery is crucial for improving the economic viability of the recycling process. Among the four selective lithium extraction methods mentioned above, mechanical activation, high-temperature phase transformation, and electrochemical methods can achieve efficient lithium recovery from spent LIBs. However, their industrial application is currently limited by factors such as small equipment scale and high energy consumption. Oxidant-based oxidative lithium extraction technology is more suitable for large-scale industrial production due to its short process flow, high selectivity, and simple operation, and it is currently the dominant industrialized process.

4.2.1.2 Recycling and Utilization of Iron Phosphorus Residue After Lithium Extraction

$FePO_4$ precipitate is typically considered a low-value by-product (with a market price of approximately 10,325 CNY/ton). However, its application value decreases further when impurity levels exceed regulatory limits. Moreover, phosphorus recovery requires additional processing steps and offers limited economic returns, leading most enterprises to prioritize lithium extraction at the expense of phosphorus resource utilization. To address this challenge, CITIC Guoan implemented a mixed acid system consisting of sulfuric acid and citric acid (with H_2SO_4 concentration maintained below 1.5 mol/L). In this system, citric acid is used to complex Fe^{3+} and inhibit its dissolution, reducing acid consumption and mitigating the impact of iron impurities. While maintaining the lithium leaching efficiency of 92%, the iron dissolution rate was effectively suppressed to below 5%. Hydrometallurgy has become the preferred process for LFP battery recycling due to its high lithium recovery rate and the potential for direct material return to the cathode, particularly suitable for spent electrodes with a single composition and thorough pretreatment.

Liu et al. directly leached $FePO_4$ slag with hydrochloric acid, followed by filtration and pH adjustment of the leachate using an alkaline agent to precipitate iron, obtaining $FePO_4{\cdot}2H_2O$ precipitate. Under the conditions of 3 mol/L hydrochloric acid concentration, a liquid-to-solid ratio of 15 mL/g, a reaction time of 4 hours and leaching temperature of 30 °C, the iron phosphate recovery efficiency reached 92.4%.[31] Yang et al. employed a method involving low-concentration acid leaching followed by high-temperature carbon removal to recover $FePO_4$ from lithium extraction slag. The process and mechanism are illustrated in Figure 4.4a. Under the conditions of 1.5 mol/L phosphoric acid concentration, H_3PO_4/HCl molar ratio of 3:1, a liquid-to-solid ratio of 10 mL/g, a leaching temperature of 90 °C and leaching time of 3 hours, the leaching efficiencies of impurity elements such as Ni, Cu, and Mn reached 99.53, 98.08, and 97.65%, respectively, whereas those of the major elements Fe and P were only 0.3 and 1.6%, respectively. Subsequently, the resulting leaching residue was calcined at 600 °C for 1 hour to remove residual carbon. The electrochemical performance of the regenerated LFP was 151.4 mAh/g at 0.2 °C, with a capacity retention rate of 94.4% after 100 cycles.[32]

Wang et al. proposed a mechanochemical process for the recovery of $FePO_4$ slag, with the process and mechanism illustrated in Figure 4.4b. During this process, $FePO_4$ slag is pretreated by ball milling and subsequently leached with sulfuric acid solution. Under the optimized ball milling conditions (with a mass ratio of stainless-steel balls to material to water of 2:1:2.5 and a milling time of 20 minutes), the maximum particle size of $FePO_4$ slag decreased from 34.265 to 13.102 μm, while the specific surface area increased from 11.41 to 13.74 m^2/g. Under the optimized leaching conditions (temperature of 333 K, acid-to-material ratio of 0.46 mL/g, liquid-to-solid ratio of 5:1 mL/g, and stirring speed of 600 rpm), the iron leaching efficiency reached 98%. Kinetic studies indicate that the leaching process is governed by both diffusion and chemical reaction, with an apparent activation energy of 29 kJ/mol.[33] Zhou et al. synthesized $LiMn_{0.25}Fe_{0.75}PO_4$ (LMFP) cathode material by mechanically mixing $FePO_4$ slag, Li_2CO_3, $MnCO_3$, and glucose, followed by carbothermal reduction, with the process and mechanism illustrated in Figure 4.4c. This cathode material features Mn gradient doping and a uniformly distributed carbon coating layer, demonstrating excellent electrochemical performance. The discharge capacity at 0.1C (161.3 mAh/g), rate capability at 5C (90.1 mAh/g), cycling stability at 1C (the capacity retention rate of 95.6% after 800 cycles), and energy density (approximately 15% higher than that of LFP) are comparable to those of commercial cathode materials.[34]

Hu et al. found that, with the assistance of iron powder, low-concentration H_2SO_4 can be used to achieve efficient leaching of $FePO_4$ slag, significantly reducing

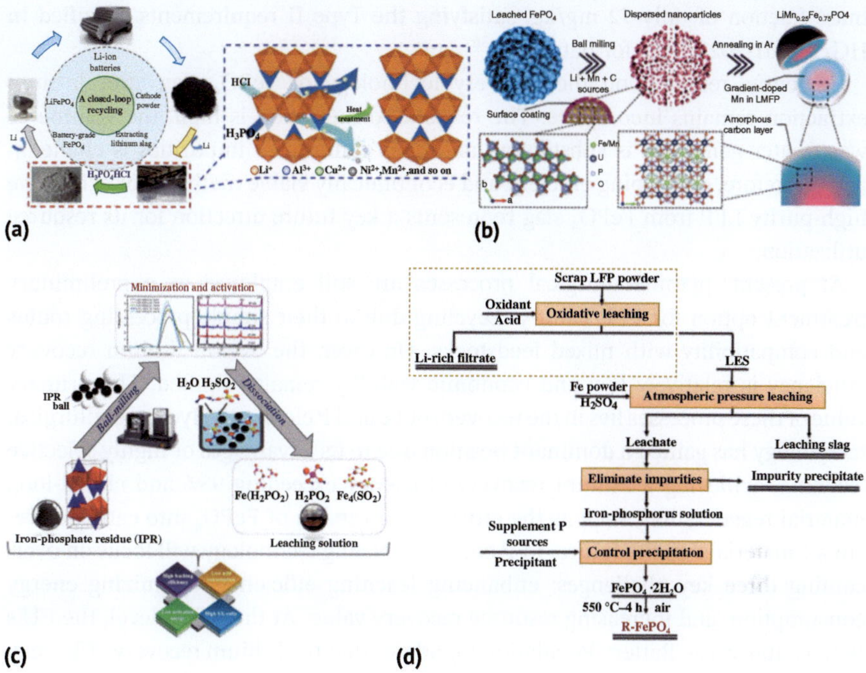

Figure 4.4 (a) The process and mechanism for preparing battery-grade $FePO_4$ from lithium-extracted LFP slag. *Source:* [32] / with permission of Elsevier. (b) Schematic illustration for the upcycling process of leached $FePO_4$ toward LMFP material. *Source:* [34] / with permission of John Wiley & Sons. (c) Sulfuric acid leaching of ball-milled activated $FePO_4$ residue after lithium extraction from spent LFP cathode powder. *Source:* [33] / with permission of Elsevier. (d) Schematic diagram of the lithium extraction slag recycling process. *Source:* [35] / with permission of Elsevier.

recovery costs, as illustrated in Figure 4.4d. Under the optimal leaching conditions (H_2SO_4 molar concentration at 1.4 times the stoichiometric requirement, Fe powder dosage at 1.5 times, liquid-to-solid ratio of 5 mL/g, reaction temperature of 60 °C, and reaction time of 150 minutes), the leaching efficiencies of Fe and P reached 94.66 and 98.23%, respectively. The discharge specific capacity of the LFP cathode material synthesized via ball milling, sand milling, and spray drying processes at 1C is 146.6 mAh/g. After 200 cycles, the capacity remains stable at 146.7 mAh/g. The mass of the regenerated R-$FePO_4$ matches that of commercial $FePO_4$.[35]

Yao et al. proposed a method for recovering $FePO_4$ from $FePO_4$ slag using nitric acid leaching. Firstly, methanol is added to the leaching solution to electrochemically reduce iron ions. Subsequently, hydrofluoric acid is introduced into the reduced solution as a precipitating agent to remove aluminum ions from the acid leaching solution. After aluminum removal, Fe^{2+} is oxidized to Fe^{3+} using hydrogen peroxide, followed by the addition of formaldehyde and heating to precipitate battery-grade $FePO_4$. The experimental results indicate that under leaching conditions of 70 °C, nitric acid concentration of 5 mol/L, and a leaching duration of 40 minutes, the Fe leaching efficiency reaches 92.7%. Following electrolytic reduction of ferrous ions and aluminum removal, the synthesized LFP exhibits an aluminum mass fraction of only 72 mg/kg, satisfying the Type II requirements specified in HG/T 4701-2021 "LFP for Batteries."[36]

Currently, research on the recovery technology of $FePO_4$ slag after lithium extraction remains incomplete. The reagent consumption is high, the volume of wastewater generated is substantial, and deep removal of impurities is challenging. Therefore, developing efficient and economically viable methods for extracting high-purity LFP from $FePO_4$ slag represents a key future direction for its resource utilization.

At present, pyrometallurgical processes are still employed as a preliminary treatment option for LFP battery recycling due to their simple processing routes and compatibility with mixed feedstocks. However, the overall lithium recovery efficiency is relatively low and economic viability remains limited. The primary value of these processes lies in the recovery of Fe and P elements. Hydrometallurgical technology has gained a dominant position due to its advantages of highly selective leaching, achieving a lithium recovery efficiency exceeding 95% and closed-loop material regeneration, such as the efficient conversion of $FePO_4$ into cathode precursor materials. In the future, LFP battery recycling technology will focus on overcoming three key challenges: enhancing leaching efficiency, optimizing energy consumption, and increasing resource recovery value. At the policy level, the EU's Battery and Waste Battery Regulation mandates that the lithium recovery efficiency exceed 90% and the recycling efficiency of spent battery materials surpass 70% by 2030, effectively phasing out pyrometallurgical processes. In contrast, emerging markets in Southeast Asia, such as Indonesia and Vietnam, permit the use of pyrometallurgy as a transitional technology due to their relatively lenient environmental regulations and lower investment requirements for pyrometallurgical infrastructure. China, on the other hand, adopts a "gradient adaptation" strategy. It prioritizes hydrometallurgical processing for high-value battery waste, such as cathode and anode black mass, while permitting pyrometallurgical pretreatment for low-grade mixed waste prior to purification. This approach balances resource efficiency with practical feasibility.

4.2.2 Analysis of Hydrometallurgical Recycling for LFP

The hydrometallurgical recycling process recovers valuable elements from LFP batteries through acid leaching and chemical precipitation, but its technical limitations primarily include high reagent consumption, significant wastewater treatment requirements, and structural performance deficiencies in the regenerated materials.

Firstly, the hydrometallurgical process requires extensive use of sulfuric or hydrochloric acid to dissolve LFP, along with hydrogen peroxide and other oxidants to facilitate the oxidation of iron ions. The highly concentrated acidic wastewater generated from LFP treatment contains iron, phosphorus, and lithium ions, requiring multistage treatment through neutralization, precipitation, and filtration. Moreover, the phosphorus-containing sludge generated during the neutralization process is difficult to utilize in a resource-efficient manner, and the salts present in the wastewater, such as Na_2SO_4, are prone to cause soil salinization, thereby further increasing the cost of environmental management. Secondly, hydrometallurgical processes are highly sensitive to microstructural changes induced by material aging, such as particle microcracks. The microcracks and surface passivation layers on LFP particles induced by cyclic aging reduce leaching efficiency, necessitating prolonged reaction times or elevated temperatures and thereby increasing reagent and energy consumption. In addition, the hydrometallurgical process struggles to completely remove aluminum and copper impurities present in the cathode material. These impurities can interfere with the subsequent precipitation process, leading to insufficient purity of the regenerated material and adversely affecting its performance. For instance, regenerated LFP containing aluminum is prone to lattice defects and exhibits significantly lower rate capability and cycle life compared to the original material.

4.2.3 Application of Hydrometallurgical Recycling for LCO

Hydrometallurgical processes play a central role in the recycling of LCO batteries. Their "leaching-separation-purification" technical framework enables selective extraction of cobalt while simultaneously recovering other valuable metals such as lithium, aluminum, and copper, thereby fully aligning with the metal-rich, high-value composition characteristic of LCO batteries. In the recycling process, LCO black powder or electrode sheets, following crushing and disintegration, undergo pretreatment steps such as thermal decomposition to remove residual electrolytes, and magnetic separation or sieving to eliminate separators and casing fragments. The pretreated material is then introduced into the acid leaching system, where sulfuric acid or hydrochloric acid serves as the primary leaching agent, typically combined with the reducing agent such as hydrogen peroxide or sodium bisulfite ($NaHSO_3$) to promote the efficient dissolution of cobalt and lithium from the spent LCO. The cobalt leaching efficiency can exceed 98%. In comparison with pyrometallurgical high-temperature reduction smelting, which recovers only metal alloys and achieves a cobalt recovery efficiency of 75–85%, the hydrometallurgical process significantly enhances both the direct recovery efficiency of valuable metals and its adaptability to diverse feedstock materials. The core advantages of hydrometallurgical processes in LCO battery recycling are clearly demonstrated in three key aspects: high efficiency, operational flexibility, and environmental sustainability. Cobalt, as the core value component—with an international market price of approximately $35,000 per ton—can alone account for over 70% of the total value recovered in the hydrometallurgical recycling process. The high-purity cobalt sulfate or cobalt oxide produced can be directly utilized as the precursor for battery cathode materials, avoiding the resource losses and additional costs associated with secondary refining of low-grade Co-Fe alloys generated through pyrometallurgical processes. Second, hydrometallurgical processes exhibit high raw material compatibility. They are not only suitable for processing single-composition LCO battery black powder but can also recover nickel, manganese, and other associated metals through dynamic adjustment of the extractant system—for instance, by optimizing the ratio of Cyanex 272 to D2EHPA. This capability provides a comprehensive solution for recycling spent batteries containing mixed LCO and ternary cathode materials, effectively demonstrating the process's technical adaptability in real-world industrial settings where diverse battery types are co-collected within electric vehicle battery packs. In industrial practice, GEM Company disassembles and sorts retired batteries, followed by efficient dissolution through the acid leaching process. Subsequently, specific extractants are employed to sequentially purify cobalt and lithium, ultimately yielding battery-grade Co_3O_4 (purity ≥ 99.5%) and Li_2CO_3 (purity ≥ 99.9%). The integrated waste gas and wastewater treatment system, achieving the removal efficiency of over 99.9%, ensures compliance with environmental protection standards. This production line processes 50,000 tons of black powder annually, with a comprehensive cost 30% lower than that of pyrometallurgical processes. The products are directly supplied to battery material recycling, enabling a "waste-to-raw material" closed-loop system and advancing resource circularity.

In the laboratory, researchers are continuously advancing process innovations to improve recovery efficiency and reduce regeneration costs. Current approaches include the use of inorganic acids, organic acids, and ionic liquids.

Inorganic acids are strong acids capable of enabling rapid leaching reactions and achieving high metal recovery efficiency from spent LIBs. In addition, strong acids with high dissociation constants, when combined with reducing agents or halides, can enhance the leaching efficiency. Acid leaching can be performed directly using either the pretreated electrode material or the residual solid feed obtained after alkaline leaching. The leaching reagents commonly employed include sulfuric acid, hydrochloric acid, and nitric acid. The efficiency of acid leaching is further enhanced by the addition of hydrogen peroxide as an oxidant. H_2SO_4 is the most commonly used inorganic acid for leaching valuable metals from spent materials. Sulfuric acid has been widely studied, with numerous research groups reporting varied results on the efficient leaching of critical metals such as Li and Co from spent LIBs. The addition of H_2O_2 is commonly employed to enhance leaching efficiency and promote reaction kinetics. In the study by Swain et al., the presence of H_2O_2 was found to shorten the leaching time.[37] Sun and Qiu reported that under conditions of 80 °C, 2 mol/L (hereinafter referred to as M) sulfuric acid solution, and a solid-to-liquid ratio of 50 g/L for 60 minutes, the recovery efficiencies of lithium and cobalt reached 99%, as shown in Eq. (1).[38]

$$LiCoO_2 + 3H_2SO_4 + H_2O_2 \rightarrow Li_2SO_4 + 2CoSO_4 + 4H_2O + 1/(2O_2) \quad (1)$$

Ferreira et al. reported that Co_3O_4 remains in the residual solid after the leaching of $LiCoO_2$ (hereinafter referred to as LCO) batteries using H_2SO_4. Consequently, an excess of H_2SO_4 is necessary to convert Co_3O_4 into soluble $CoSO_4$ through the following sequential reactions, as given in Eqs. (2) and (3).

$$4LiCoO_2 + 3H_2SO_4 \rightarrow Co_3O_4 + 2Li_2SO_4 + CoSO_4 + 3H_2O \quad (2)$$

$$Co_3O_4 + 3H_2SO_4 \rightarrow 2Li_2SO_4 + 3H_2O + 3CoSO_4 + 1/(2O_2) \quad (3)$$

The overall reaction equation is given in Eq. (4).

$$4LiCoO_2 + 6H_2SO_4 \rightarrow O_2 + 2Li_2SO_4 + 4CoSO_4 + 6H_2O \quad (4)$$

In this study, the recovery rates of Li and Co using H_2SO_4 were reported to be 100 and 97%, respectively.[39] Wang et al. investigated the influence of $Na_2S_2O_3$ and H_2SO_4 concentrations on the leaching efficiencies of Co, Cu, and Li, and found that increasing the $Na_2S_2O_3$ concentration led to a decrease in the leaching rates of these metals. They reported the optimal leaching conditions as follows: 3 M H_2SO_4, a liquid-to-solid ratio of 15:1, 0.25 M $Na_2S_2O_3$, 90 °C, and a duration of 3 hours. Under these optimized conditions, the recovery rates of lithium and cobalt reached 99.71 and 99.95%, respectively. The leaching reaction is shown in Eq. (5).[40]

$$LiCoO_2 + 3H_2SO_4 + Na_2S_2O_3 \rightarrow Li_2SO_4 + 2CoSO_4 + NaSO_4 + S + 3H_2O \quad (5)$$

Takacova et al. reported the optimal conditions for leaching electrode waste using HCl solution: 2 M HCl, 60–80 °C, and 90 minutes, achieving nearly 100% extraction of lithium and cobalt. They found that HCl leaching was more effective than H_2SO_4. The chemical reaction between lithium and hydrochloric acid is given in Eq. (6).[41]

$$2LiCoO_2 + 8HCl \rightarrow 2LiCl + 2CoCl_4 + 4H_2O + Cl_2 \quad (6)$$

Shuva et al. conducted the kinetic study on hydrochloric acid leaching for the efficient recovery of lithium and cobalt from spent LCO cathodes. The results show that when the reaction is conducted at 80 °C for 60 minutes, the leaching efficiencies of lithium and cobalt reach 83.7 and 83%, respectively. Kinetic analysis shows that the leaching of lithium (with an activation energy of 23.83 kJ/mol) and cobalt (27.72 kJ/mol) follows the chemical reaction-controlled model. This study provides a theoretical foundation for the efficient recovery of battery metals in the hydrochloric acid system, and the optimized conditions enable a balance between leaching efficiency and energy consumption control.[42]

Similarly, it was found that the HNO_3 system is effective for leaching Li and Co from spent LIBs. Castillo et al. achieved 100% leaching efficiency of lithium from spent LIBs. Few research groups have investigated the effect of systematically adding H_2O_2 to HNO_3, and have found that introducing this additive increases the number of chemical reaction steps, thereby reducing leaching time.[43] For lithium leaching, the chemical reaction control model, diffusion-controlled model, or shrinking core model is applicable. For cobalt leaching, the ash diffusion-controlled model for constant-size spherical particles provides the best fit. The activation energies for the leaching of Li and Co are 32.4 and 59.81 kJ/mol, respectively.[44] Chen et al. employed the phosphoric acid leaching-precipitation process to recover cobalt and lithium from spent LCO cathodes. Using H_3PO_4 as both the leaching and precipitating agent, under optimal conditions, the leaching efficiencies of cobalt and lithium exceeded 99% within 60 minutes. Cobalt precipitates as $Co_3(PO_4)_2$ with a purity of 97.1%, while lithium remains in the liquid phase. This process enables efficient metal separation under mild conditions, simplifies the traditional multistep procedure, and integrates economic efficiency with environmental sustainability.[45] Zheng et al. employed formic acid leaching for the treatment of spent LCO. Under optimal conditions, the leaching efficiencies of lithium and cobalt reached 99.9 and 99.96%, respectively, within 20 minutes. After stepwise precipitation of the leaching solution, cobalt is recovered as hydroxide, and lithium is obtained in the form of lithium fluoride (LiF) with a purity of 99% via NaF precipitation. The entire process enables efficient metal separation under mild conditions, offering a new approach for the green recovery of LiF.[46]

Golmohammadzadeh et al. conducted a systematic investigation into the process optimization and kinetics of leaching cobalt and lithium from spent LIBs using organic acids. The process parameters were optimized through the response surface methodology, and ultrasonic assistance was employed to enhance metal leaching efficiency. The results indicated that temperature exerts the most significant influence on the leaching performance. Under the optimal conditions, the recovery efficiencies of lithium and cobalt reached 92 and 81%, respectively. The kinetics of cobalt leaching is governed by surface chemical reactions at temperatures below 45 °C. According to the shrinking core model analysis, citric acid exhibits a superior leaching performance compared to other reagents due to its strong chelating ability. This study provides a kinetic foundation for metal recovery in organic acid-based systems.[47]

In the recycling of LCO batteries, pyrometallurgical and hydrometallurgical processes differ in their environmental pollution risks and policy support considerations. The hydrometallurgical process generates strong acid wastewater (e.g., sulfuric acid leaching solution), residual organic extractants (e.g., P507), and nitrogen oxide emissions, leading to the formation of a potential chemical pollution chain. Environmental impacts are mitigated through waste acid recovery via membrane technologies—such as BASF's system achieving an 80% recovery rate—and solvent-free extraction techniques. The pyrometallurgical process is prone to generating HF off-gas (from PVDF decomposition) and heavy metal-containing slag, and relies on flue gas purification and carbon emission reduction technologies to mitigate high-temperature pollution. In terms of policy support, China provides subsidies for waste heat power generation from pyrometallurgical processes at a rate of 0.1 CNY/kWh to reduce energy consumption, while hydrometallurgical processes benefit from 30% subsidy on waste acid treatment equipment. Overall, pyrometallurgy dominates the recycling of power batteries due to its high processing efficiency, whereas hydrometallurgy prevails in the consumer battery market owing to its ability to produce high-purity cobalt (>99.95%). The "combined pyrometallurgical and hydrometallurgical" approach promoted in China and Europe—featuring pyrometallurgical pretreatment followed by hydrometallurgical refining—integrates the advantages of both methods and achieves the dual objectives of high efficiency and environmental sustainability.

4.2.4 Technical Analysis of the Hydrometallurgical Process for LCO Recovery

The hydrometallurgical process recovers valuable metals from LCO through chemical leaching and separation techniques. However, its technical limitations are primarily associated with high reagent consumption, complex wastewater treatment requirements, and performance deficiencies in the regenerated materials.

Although the hydrometallurgical process can efficiently dissolve cobalt, lithium, and other elements from LCO through acid leaching, their dissolution, complexation, and extraction behaviors are highly similar. For instance, Co^{3+} and Li^{+} often coexist in solution after leaching. Traditional extractants struggle to selectively target individual ions. To meet stringent purity requirements, multi-stage operations are necessary to enable the stepwise recovery of the target metal. During this process, cobalt may be lost due to secondary dissolution, whereas lithium is prone to being co-extracted into the impurity phase by chelating agents. Consequently, simultaneous enhancement of the recovery efficiencies of both metals becomes challenging. In addition, the vigorous reaction of aluminum foil current collectors in acidic environments further increases recycling costs. The rapid corrosion of aluminum not only consumes excessive amounts of acidic reagents but also releases Al^{3+}, which competes with cobalt for the active sites of the extractant and can form gel-like aluminum hydroxide that clogs pipelines, necessitating the use of additional centrifugation or flocculation units to maintain system stability. Secondly, the issue of sensitivity to organic residues in the hydrometallurgical process is equally prominent. The hydrofluoric acid generated from the decomposition of fluorides in the electrolyte upon contact with water is highly corrosive to equipment. Meanwhile, residual PVDF binder can coat LCO particles, impeding full contact with the acidic solution. To mitigate this issue, high-temperature pretreatment or strong oxidants are required, increasing energy consumption and reagent costs. Meanwhile, impurity metals such as copper and iron introduced during the disassembly of LCO batteries compete with target metals for dissolution in the acid leaching stage. In the subsequent impurity removal process, iron ions require frequent pH adjustments to enable stepwise precipitation. This not only prolongs the reaction cycle but may also interfere with the purity of cobalt precipitates, thereby compromising the electrochemical performance of the regenerated materials. Finally, when dealing with retired LCO batteries of varying compositions, it is necessary to continuously optimize reaction parameters to maintain separation efficiency; however, the practical challenges associated with dynamically adjusting acid-base concentration or redox potential are considerable. Moreover, the high-salt wastewater and heavy metal-bearing waste residues generated during the processing flow require additional facilities for neutralization, adsorption, and solidification, which further compresses the profit margin of metal recovery and limits the large-scale applicability of the hydrometallurgical process in LCO battery recycling.

4.2.5 Application of Hydrometallurgical Recycling for Ternary Battery

In the recycling of NCM batteries, hydrometallurgical processes also play a crucial role. The core of this approach involves the targeted recovery of high-value metals such as lithium, cobalt, and nickel through sequential steps including acid or alkali leaching, selective separation, and chemical precipitation. The process typically consists of three major stages: pretreatment (including discharge, crushing, and screening), acid leaching of active materials, and metal separation and purification. Firstly, the cathode black mass or crushed material obtained after disassembly undergoes reductive leaching in the mixed solution. Under optimal conditions, the metal leaching efficiency exceeds 95%. Subsequently, the metal ions are selectively separated through multiple stages of solvent extraction, coprecipitation, or electrolysis. Taking the resynthesis of the cobalt phosphate precursor as an example, pH adjustment enables the selective precipitation of cobalt, while nickel and manganese are further purified through chelating resin adsorption. For lithium recovery, the industrial sodium carbonate precipitation method is widely employed. Following high-temperature concentration of the leaching solution, saturated sodium carbonate is added to precipitate lithium carbonate with a purity of 99.2–99.8%. Among current mainstream technologies, China Bangpu Company achieves a comprehensive metal recovery rate of over 98% through the "synergistic acid leaching–extraction diversion" process; however, additional processing equipment is still required for manganese enrichment. Based on the types of leaching agents employed in the leaching process, it can be broadly classified into inorganic acid leaching, organic acid leaching, and ammonia leaching. Table 4.2 summarizes hydrometallurgical process methods for the treatment of retired ternary LIB materials reported in the literature.

Zou et al. leached mixed cathode materials composed of LCO, $LiMn_2O_4$, $LiNi_{0.33}Mn_{0.33}Co_{0.33}O_2$, and LFP using a sulfuric acid–hydrogen peroxide system. Under optimal conditions of temperature, time, and other parameters, $Fe(OH)_3$ and Ni/Co/Mn hydroxides were sequentially precipitated through pH regulation, followed by the recovery of Li_2CO_3. Ni-Co-Mn hydroxide and Li_2CO_3 were ball-milled and calcined to synthesize $LiNi_{0.33}Mn_{0.33}Co_{0.33}O_2$. After 50 cycles at 0.5 °C, the capacity retention reached 82.4%. The overall process enables direct synthesis of ternary materials without requiring metal separation, thereby reducing costs. It is compatible with multiple cathode chemistries, facilitating closed-loop resource utilization.[61] Li et al. developed a lactic acid-based leaching process for the recovery of valuable metals from spent $LiNi_{1/3}Co_{1/3}Mn_{1/3}O_2$ cathode materials. Under optimal conditions, the leaching efficiency of NCM reaches 97.7–98.9%. The kinetic analysis conforms to the Avrami equation, with an activation energy ranging from 62.8 to 70.6 kJ/mol, indicating that the overall reaction is controlled by surface chemistry. The leaching solution was converted into cathode material via the sol-gel method. After 100 cycles at 0.5 °C, the discharge capacity reached 138.2 mAh/g with a capacity retention of 96%, comparable to that of the newly synthesized material.[62] Gao et al. selectively leached spent $LiNi_{1/3}Co_{1/3}Mn_{1/3}O_2$ using acetic acid. Under optimal conditions, the leaching efficiencies of Co, Li, Mn, and Ni reached 93.6, 99.97, 96.3, and 92.7%, respectively, whereas the leaching efficiency of Al was only 2.36%. Kinetic analysis indicates that the reductant shifts the rate-controlling step from diffusion through the residual layer to the surface chemical reaction, thereby reducing the activation energy to 41.2–42.3 kJ/mol. Finally, after pH adjustment and precipitation of the leachate, Li_2CO_3 with a purity of 99.93% was obtained, and the overall metal recovery rate exceeded 90%.[63]

The process flow of the current fully hydrometallurgical recycling method for treating spent ternary LIBs in the industry is illustrated in Figure 4.5, as follows: (1)

Table 4.2 Hydrometallurgical process for spent ternary LIB materials.

Raw material	Leaching agent with reductant	Temperature (°C)	Time (min)	Liquid-to-solid ratio (mL/g)	Leaching efficiency (%)	References
Inorganic acid						
NCM	4 M HCl	80	60	50	Li, Ni, Co, Mn > 99	2
NCM	1 M H_2SO_4	95	240	50	Li: 93.4, Ni: 96.3, Co: 66.2, Mn: 50.2	48
NCM	1 M H_2SO_4 + 1 vol.% H_2O_2	40	60	40	Li, Ni, Co, Mn: 99.7	49
NCM	1 M H_2SO_4 + 5 vol.% H_2O_2	95	240	50	Li: 94.5, Ni: 96.4, Co: 79.2, Mn: 84.6	50
NCM	4 M H_2SO_4 + 2 M H_2O_2	90	120	125	Ni: 99, Co: 99, Mn: 97	51
NCM	1 M H_2SO_4 + 0.075 M $NaHSO_3$	95	240	20	Li: 96.7, Ni: 96.4, Co: 91.6, Mn: 87.9	52
NCM	2.5 M H_2SO_4 + 0.8 M NH_4Cl	80	60	100	Li: 99, Ni: 98, Co: 98, Mn: 97	53
NCM	0.8 M H_2SO_4 + 0.075 M Na_2SO_3	70	90	10	Li: 99, Ni: 98.6, Co: 97.9, Mn: 99.1	54
Organic acid						
NCM	1 M citric acid + 12 vol.% H_2O_2	60	40	12.5	Li, Ni, Co, Mn > 98	55
NCM	1 M acetic acid + 6 vol.% H_2O_2	70	60	50	Li: 98.4, Ni: 97.3, Co: 97.7, Mn: 97.1	56
NCM	1.5 M tartaric acid + 2 vol.% H_2O_2	80	30	40	Li: 99.7, Ni: 6.5, Co: 2.5, Mn: 3.9	57

(Continued)

Table 4.2 (Continued)

Raw material	Leaching agent with reductant	Temperature (°C)	Time (min)	Liquid-to-solid ratio (mL/g)	Leaching efficiency (%)	References
Ammonia leaching						
NCM	1 M $(NH_4)_2CO_3$ + 4 M NH_3 + 0.3 M Na_2SO_3	80	5.0	100	Li: 79.1, Ni: 85.3, Co: 86.4	[58]
NCM	1.5 M $(NH_4)_2SO_3$ + 4 M NH_3 + 0.5 M Na_2SO_3	80	5.0	100	Li: 95.3, Ni: 89.8, Co: 80.7	[59]
NCM	3 M $(NH_4)_2SO_4$ + 1.5 M $(NH_4)_2SO_3$	180	2.0	12	Li: 98, Ni: 98, Co: 81	[60]

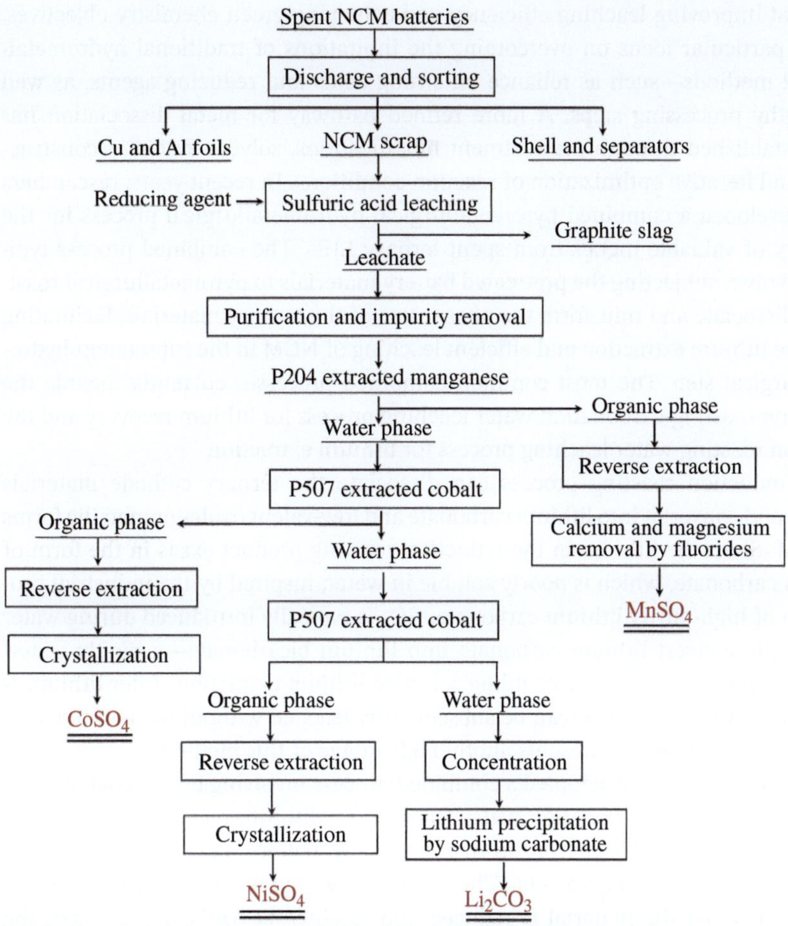

Figure 4.5 Schematic flowchart of the fully hydrometallurgical process for recycling ternary LIBs.

Spent ternary LIBs are first discharged and then subjected to physical sorting, yielding ternary cathode black mass, mixtures of copper and aluminum foils, battery casings and separators, among others. (2) The black powder is leached with sulfuric acid. During the acid leaching process, reducing agents such as hydrogen peroxide or sodium sulfite are commonly added. The leaching residue primarily consists of a mixture of anode graphite and the conductive agent acetylene black. (3) After purification of the leaching solution, manganese, cobalt, and nickel are separated via solvent extraction to yield sulfate products. (4) The lithium concentration in the raffinate is very low (<2 g/L), whereas the concentration of coexisting sodium or ammonium ions is considerably high. Therefore, in industrial practice, lithium is typically recovered by concentration followed by the addition of sodium carbonate to precipitate it as lithium carbonate.

In response to the low lithium recovery efficiency (approximately 75%) in the fully hydrometallurgical process, researchers have implemented diverse innovations aimed at improving leaching efficiency and achieving green chemistry objectives, with a particular focus on overcoming the limitations of traditional hydrometallurgical methods—such as reliance on strong acids and reducing agents, as well as lengthy processing steps. A more refined pathway for metal dissociation has been established through pretreatment modifications, solvent system reconstructions, and iterative optimization of reaction conditions. In recent years, researchers have developed a combined pyrometallurgical-hydrometallurgical process for the recovery of valuable metals from spent ternary LIBs. The combined process typically involves subjecting the pretreated battery materials to pyrometallurgical roasting to dissociate and transform the phase state of the ternary materials, facilitating selective lithium extraction and efficient leaching of NCM in the subsequent hydrometallurgical step. The most common combined processes currently include the reduction roasting-carbonation water leaching process for lithium recovery and the sulfation roasting-water leaching process for lithium extraction.

The reduction roasting process can dissociate the ternary cathode materials (NCM) and convert it into lithium carbonate and low-valent oxides or metallic forms of NCM. Since the lithium in the reduction roasting product exists in the form of lithium carbonate, which is poorly soluble in water, inspired by the industrial production of high-purity lithium carbonate, CO_2 is typically introduced during water leaching to convert lithium carbonate into lithium bicarbonate—a highly water-soluble compound—thereby enabling selective lithium extraction. After lithium is leached by water and NCM can be subsequently leached without the addition of a reducing agent, which represents another advantage of this method.

For instance, Hu et al. adopted a combined process involving brown coal-assisted reduction roasting, carbonation water leaching for lithium recovery, and sulfuric acid leaching for NCM recovery to treat spent LIB battery materials. The specific flowchart is shown in Figure 4.6a. The results show that when the carbon content is 19.9 wt.% and the material is reduced and calcined at 650 °C for 3 hours, the $LiNi_xCo_yMn_zO_2$ cathode material dissociates and is converted into Li_2CO_3, NCM. Subsequently, under conditions of 20 mL/min CO_2 flow rate, liquid-to-solid ratio of 10 mL/g, extraction duration of 2 hours and room temperature, lithium is selectively leached from the calcined product via carbonation water leaching. A high-purity lithium carbonate product was obtained by direct evaporation of the filtrate. Subsequently, the residual solid was leached with 3.5 mol/L H_2SO_4 solution at 85 °C for 3 hours. The experimental results indicated that 84.7% of Li and over 99% of NCM were recovered from spent ternary LIB materials using this method.[64]

Yang et al. employed methane as a reducing agent to perform reduction roasting on spent ternary LIB materials and adopted the combined carbonation water leaching and acid leaching approach for the extraction of valuable metals. They found that the presence of aluminum in the cathode material significantly affects the lithium leaching efficiency. Figure 4.6b presents the phase transformation mechanism diagram of the cathode material during the methane reduction calcination process. Firstly, in the presence of aluminum in the cathode material and at the optimal calcination temperature of 600 °C, part of the lithium released from the decomposition of $LiNi_xCo_yMn_zO_2$ combines with aluminum to form lithium

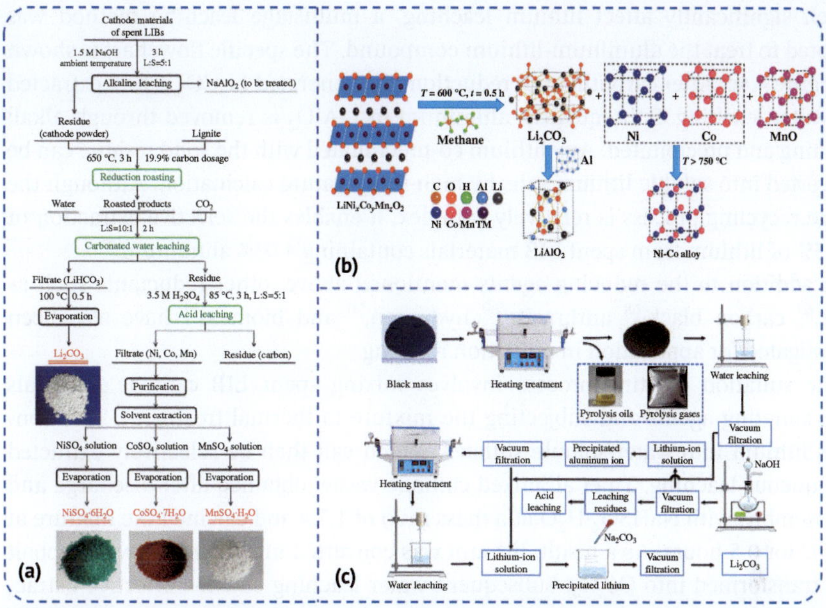

Figure 4.6 (a) A flowsheet of the approach for the recovery of valuable metals from spent LIBs. *Source:* [64] / with permission of Elsevier. (b) Illustration of the phase transformation mechanism of the spent $LiNi_xCo_yMn_zO_2$ cathode material in the reduction roasting with methane. *Source:* [65] / with permission of Elsevier. (c) Research flowchart of carbothermal reduction combined with multistage leaching. *Source:* [66] / with permission of Elsevier.

aluminate. However, after pre-alkali leaching to remove aluminum, lithium is fully converted into lithium carbonate. In addition, the reduction process of NCM is not affected by aluminum, and all three metals can be converted into acid-soluble forms: NCM. However, as the calcination temperature increases (>750 °C), nickel and cobalt tend to form alloys that are difficult to leach with acid. The specific reaction equations for the reduction of cathode materials in the presence and absence of aluminum are as follows:

$$LiNi_xCo_yMn_zO_2 + CH_4 \xrightarrow{600\,^\circ C} Li_2CO_3 + Ni + Co + MnO + CO_2 + H_2O \tag{7}$$

$$\begin{aligned} LiNi_xCo_yMn_zO_2 + CH_4 + Al \xrightarrow{600\,^\circ C} & Li_2CO_3 + LiAlO_2 + Ni + Co + MnO \\ & + CO_2 + H_2O \end{aligned} \tag{8}$$

After the pre-alkali leaching method was applied to selectively remove aluminum, the lithium leaching rate increased from 70 to 88%. In the subsequent acid leaching process without the reducing agent, the leaching efficiencies of NCM all exceeded 98%. Therefore, efficient recovery of valuable metals depends on precise control of the product phase during the reduction roasting process.[65]

Zhang et al. directly used anode graphite recovered from spent LIBs as the reducing agent. After carbothermal reduction roasting, metallic lithium was selectively extracted. Due to the presence of aluminum impurities in the raw materials, which significantly affect lithium leaching, a multistage leaching method was adopted to treat the aluminum-lithium compound. The specific flowchart is shown in Figure 4.6c. After carbothermal reduction, the generated Li_2CO_3 is first extracted via water leaching. Subsequently, aluminum in $LiAlO_2$ is removed through alkali leaching and precipitated. Any lithium co-precipitated with the solid residue can be converted into soluble lithium salts by high-temperature calcination. Although the entire recycling process is relatively complex, it enables the selective extraction of 87.15% of lithium from spent LIB materials containing 4.03% aluminum.[66]

In addition to the reducing agents mentioned above, other reductants such as coke,[67] carbon black,[68] anthracite,[69] hydrogen,[70] and biomass[71] have also been investigated for application in reduction roasting.

The sulfation roasting process involves mixing spent LIB cathode materials with sulfating agents and subjecting the mixture to thermal treatment. This converts lithium into water-soluble sulfates, which can then be selectively extracted via aqueous leaching. Yu et al. mixed cathode waste, obtained after discharge and disassembly, with $NaHSO_4{\cdot}H_2O$ at a mass ratio of 1:1.4 and calcined the mixture at 600 °C for 0.5 hours. As a result, lithium was converted into $LiNaSO_4$, while cobalt was transformed into Co_3O_4. Subsequent water leaching enabled selective extraction of lithium, with cobalt remaining in the residue.[72] The reaction equations representing the phase transformations during the entire roasting temperature ramp are as follows:

$$
\begin{aligned}
(\mathrm{NaHSO_4 \cdot H_2O} &\rightarrow \mathrm{NaHSO_4 + H_2O@2NaHSO_4} \rightarrow \mathrm{Na_2S_2O_7} \\
&+ \mathrm{H_2O@Na_2S_2O_7} \\
&\rightarrow \mathrm{Na_2SO_4 + SO_3@4LiCoO_2} + 2X\mathrm{SO_3} + X\mathrm{O_2} \\
&\rightarrow 4\,\mathrm{Li_{(1-x)}CoO_2} + 2X\mathrm{Li_2SO_4@Li_{(1-x)}CoO_2} \\
&\rightarrow (1-x)\,\mathrm{Li_2O + Co_3O_4 + O_2@Li_2O + SO_3} \\
&\rightarrow \mathrm{Li_2SO_4@Li_2SO_4 + Na_2SO_4} \rightarrow 2\,\mathrm{LiNaSO_4})
\end{aligned}
$$

However, the sodium sulfate roasting process requires strict control of the SO_3 partial pressure to ensure complete phase transformation, and the high concentration of Na^+ in the aqueous leachate from the roasted product may adversely affect lithium recovery efficiency.

Some researchers have also utilized solid waste green vitriol as the sulfating agent to develop a green recycling process for the selective separation of lithium metal from spent LMO batteries, as illustrated in Figure 4.7a. This process is environmentally friendly, with no gaseous SO_2 pollutant generated throughout the entire procedure. Under optimal experimental conditions, 99.50% of lithium is selectively extracted in the form of lithium sulfate, while manganese is separated as an oxide. The final Li_2CO_3 product achieves a purity of up to 99.77%. This study presents a novel technical approach for the environmentally friendly recycling of spent LMO batteries.[73]

Li et al. proposed the use of sulfuric acid as the sulfating agent instead of sulfate-based reagents. Figures 4.7b, c present the process flow diagrams for the sulfation roasting-water leaching treatment of retired LCO and NCM batteries, respectively. Taking spent NCM materials as an example, the spent cathode is mixed with concentrated sulfuric acid at a molar ratio of 2:1 and calcined at 800 °C for 2 hours. The overall sulfation transformation process can be divided into two stages: firstly, at

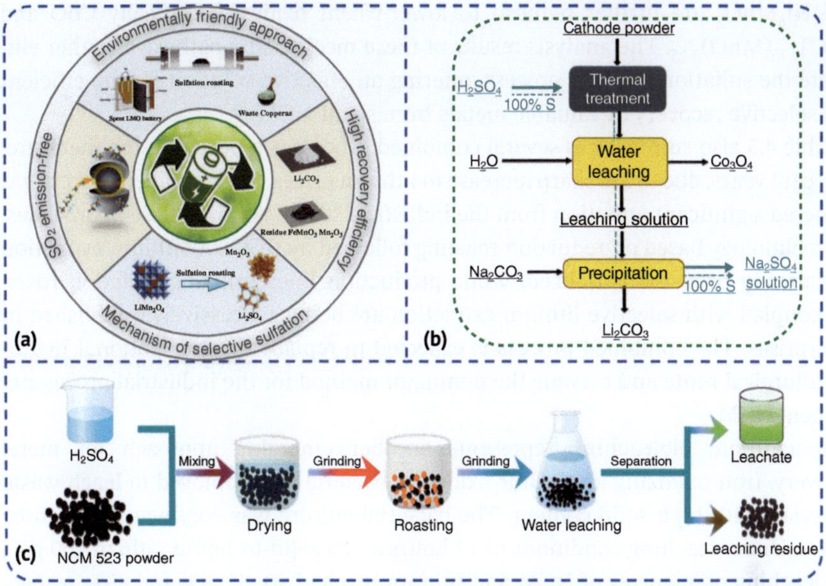

Figure 4.7 (a) Illustration of selective recovery of lithium from spent LIB by an emission-free sulfation roasting strategy. (b) Flow sheet for valuable metals recycling from the spent $LiCoO_2$ material. (c) Flow diagram for the sulfation roasting-water leaching process.

room temperature, mixing the cathode material with sulfuric acid triggers a reaction that partially disrupts the structure, during which NCM is reduced to varying extents to the divalent state, forming sulfate species ($NiSO_4$, $CoSO_4$, and $MnSO_4$). As the temperature increases continuously, lithium ions are progressively released from the destabilized layered structure and combine with SO_4^{2-} from the transition metal sulfates to form Li_2SO_4, while the transition metals react to yield $Ni_{0.5}Co_{0.2}Mn_{0.3}O_{1.4}$. Selective lithium extraction was achieved through simple water leaching, with the leaching efficiency of 96.92%, whereas other transition metals were barely leached.[74, 75]

Chen et al. mixed pre-treated cathode and anode mixed materials with concentrated sulfuric acid at a molar ratio of $n(H_2SO_4):n(Li) = 0.95:1$ and calcined the mixture at 550 °C for 3 hours. Afterward, lithium was converted into lithium sulfate, while NCM were transformed into their respective simple oxides. Following water leaching, the lithium leaching efficiency reached 90%, whereas that of manganese was only 10%. Subsequently, NCM could be recovered through sulfuric acid leaching. Due to the presence of carbon as a reducing agent in the battery raw materials, the transformation process during sulfation roasting primarily follows three pathways: when the roasting temperature is below 550 °C and the duration is less than 2 hours, lithium within the $LiCo_xNi_yMn_{1-x-y}O_2$ crystal structure undergoes intercalation and deintercalation to form soluble Li_2SO_4, while NCM are initially converted into $MnCo_2O_4$ and Ni_6MnO_8. In addition, when $n(H_2SO_4):n(Li) > 1$, a portion of lithium and manganese is further converted into water-soluble $Li_2Mn_2(SO_4)_3$, which compromises the selectivity of lithium extraction. When the calcination temperature exceeds 550 °C and the duration is longer than 2 hours, $MnCo_2O_4$

and Ni_6MnO_8 are further reduced to lower-valent oxides, specifically CoO and $(NiO)_{0.75}(MnO)_{0.25}$. The analysis results of these mechanistic pathways further elucidate the sulfation roasting process, offering an effective solution for the efficient and selective recovery of valuable metals from spent NCM.[76]

Table 4.3 also summarizes several combined processes reported in the literature. In recent years, due to the sharp increase in lithium prices, the combined process has garnered significant attention from the industrial sector. Among these approaches, the technology based on reduction roasting followed by selective lithium extraction has already been industrialized, while production lines utilizing sulfation roasting coupled with selective lithium extraction are being successively established in enterprises. The combined process is expected to replace the conventional hydrometallurgical route and become the dominant method for the industrial processing of spent NCM.

In addition, bioleaching represents another emerging approach for metal recovery. Iron-oxidizing and sulfur-oxidizing bacteria are employed to leach waste materials with high solid content. The bacterial culture was conducted over three cycles. Under leaching conditions of 72 hours and a solid-to-liquid ratio of 100 g/L, the leaching efficiencies of NCM and lithium reached 90, 82, 92, and 89%, respectively. Compared with traditional acid-base leaching, this method offers advantages in terms of environmental friendliness, sustainability, and cost-effectiveness. However, factors such as low leaching efficiency and slow leaching kinetics limit its broader industrial application.[80]

During the smelting process in pyrometallurgical technology, the release of metal dust (e.g., nickel dust and copper oxide) can lead to atmospheric particulate matter pollution. Furthermore, if residual heavy metals (e.g., lead and zinc) in the smelting slag are not properly solidified and disposed of, they may leach into rainwater and contaminate soil and groundwater. In addition, the incineration of plastic separators and electrolytes in pyrometallurgical processes may generate halogenated organic compounds containing chlorine or bromine, increasing the risk of forming persistent pollutants such as dioxins. At the policy support level, the EU's New Battery Regulation explicitly requires pyrometallurgical recycling facilities to be equipped with online monitoring systems for heavy metals. Meanwhile, in certain regions of China, such as Guangdong Province, subsidies ranging from 15 to 20% are provided for the upgrading of pyrometallurgical smelting furnaces, encouraging enterprises to install electrostatic precipitators and activated carbon adsorption units to mitigate emissions. The advantages of hydrometallurgy include its high lithium recovery efficiency, the direct usability of nickel and cobalt products for precursor synthesis, low reaction temperatures (typically $<100\,°C$), and the absence of air pollution risks. However, its large-scale application is hindered by complex procedures, high reagent consumption and organic phase loss in the extraction system. However, its large-scale application is hindered by complex procedures, high reagent consumption, and organic phase loss in the extraction system. Hydrometallurgy, owing to its advantages of high metal recovery purity and broad applicability, has become the mainstream technical route for NCM battery recycling. Nevertheless, the technology faces challenges in extraction reagent recycling,

Table 4.3 Combined pyrometallurgical-hydrometallurgical process for NCM recycling.

Raw material	Combustion of the doping agent	Roasting conditions	Leaching agent	Leaching conditions	Leaching efficiencies (%)	References
Reduction roasting						
NCM	Coke	650 °C, 30 minutes, coke dosage of 10%	H_2O	Room temperature, 60 minutes, L/S = 30	Li: 93.67 Ni: 93.33, Co: 98.08, Mn: 98.68	67
NCM	Hydrogen	500 °C, 15 minutes, hydrogen flow rate: 200 mL/min	H_2O	Room temperature, 10 minutes, L/S = 2	Li > 97, Ni, Co, Mn < 0.01	70
Sulfation roasting						
NCM	$(NH_4)_2SO_4$	350 °C, 90 minutes, $m((NH_4)_2SO_4):m(NCM)=4$	H_2O	60 °C, 30 minutes, L/S = 200	Li: 99.2, Ni: 99.4, Co: 98.8, Mn: 98.5	77
LCO	SO_2-O_2-Ar	700 °C, 120 minutes, 10 vol.% SO_2-1 vol.% O_2-89 vol. % Ar	H_2O	Room temperature, 60 minutes, L/S = 100	Li: 99.5, Co: 17.4	78
Chlorination roasting						
NCM	NH_4Cl	550 °C, 30 minutes, $m(NH_4Cl):m(NCM)=3$	H_2O	Atmospheric temperature, 10 minutes, L/S = 25	Li: 87, Ni, Co, Mn < 1	79

simultaneous purification of multiple metals, and achieving near-zero wastewater discharge. In terms of policy, China's Ministry of Ecology and Environment has classified sludge from the treatment of heavy metal-containing wastewater generated during hydrometallurgical recovery as "conditionally exempted" in the List of Hazardous Wastes Subject to Exemption Management, enabling compliant enterprises to reduce disposal costs through targeted incineration or resource utilization. The US Environmental Protection Agency (EPA) has funded hydrometallurgical enterprises through the Battery Recycling Innovation Fund to develop low-toxicity extractants—such as ionic liquids that replace traditional organophosphorus reagents—in order to reduce environmental toxicity.

4.2.6 Analysis of Hydrometallurgical Recovery of Ternary Cathode Materials

The hydrometallurgical recycling process recovers metals from ternary cathode materials via acid leaching and selective precipitation; however, its technical limitations are primarily associated with high reagent consumption, wastewater pollution, and suboptimal performance of the regenerated materials.

The limitations of hydrometallurgical processes in treating spent NCM batteries are closely linked to the characteristics of their core process steps. First, although metals such as NCM in ternary cathode materials can be dissolved in acidic reagents, their similar chemical properties make separation challenging. Cobalt and nickel tend to interfere with each other during conventional extraction processes, often necessitating multiple repetitions or the addition of specialized chemical reagents for separation. This not only prolongs processing time but also increases costs and reagent waste. Manganese is prone to react with other substances in the system and remains in the solution as an impurity, potentially causing crystalline structure defects in the subsequently resynthesized cathode material and adversely affecting battery performance. Second, the aluminum foil in NCM batteries rapidly corrodes in strongly acidic environments, releasing a significant amount of aluminum ions. These ions not only form gel-like substances that clog pipelines and filters but may also enter the metal leaching solution and ultimately contaminate the recycled materials. Even after multiple purification steps, trace aluminum impurities are difficult to completely eliminate. Under high-temperature processing conditions, they may accelerate lithium loss, compromising the structural stability of the recycled cathode materials and resulting in faster battery capacity degradation. In addition, fluorides generated from the decomposition of organic components such as electrolytes and binders may not only corrode equipment but also interfere with metal ion separation during the precipitation process and contaminate chemical reagents, further increasing treatment costs. Finally, different models of ternary LIBs (e.g., NCM523 or NCM811), or even batteries from the same production batch that have been discarded, may exhibit significant variations in internal composition due to differing usage histories. In hydrometallurgical processes, parameters such as acid and alkali concentrations and reaction time must be precisely controlled. However, when confronted with compositional fluctuations, flexible adjustments are often difficult to implement, leading to unstable metal recovery efficiency or substandard purity of the recycled materials. For instance, when fragments of other battery types (e.g., LFP) are mixed in, iron elements may dissolve in large quantities during the acid leaching stage, interfering with the subsequent separation of cobalt and nickel. Existing hydrometallurgical processes struggle to efficiently address such issues.

4.2.7 Application of Hydrometallurgical Recovery for Solid-state Batteries

The hydrometallurgical recycling process for solid-state batteries differs fundamentally from that of traditional LIBs, primarily due to the chemical inertness of the solid electrolyte, its compositional complexity, and its strong interfacial bonding with electrode materials. The hydrometallurgical recycling of traditional LIBs relies on the presence of the liquid electrolyte ($LiPF_6$ in organic solvent) and the polyolefin-based separator. This process enables efficient dissolution of cathode active materials, such as $LiCoO_2$ and NCM, via acid leaching, followed by metal separation through solvent extraction and precipitation. Solid-state batteries employ solid electrolytes—such as oxide LLZO, sulfide $Li_{10}GeP_2S_{12}$, or polymer PEO-LiTFSI—whose dense structure and high chemical stability significantly alter leaching kinetics, impurity dissolution pathways, and separation processes. The low solubility of solid electrolytes—such as oxides requiring concentrated nitric acid or hydrofluoric acid for dissolution—the risk of side reactions—such as sulfides releasing H_2S upon acid exposure—and the introduction of new impurity elements—such as rare earths, germanium, and sulfur—necessitate a complete redesign of the leaching system, the development of targeted separation technologies and enhanced pollution control strategies. In addition, solid-state batteries lack liquid electrolytes, eliminating the need to manage hazardous fluorides (e.g., HF) and organic solvents in conventional hydrometallurgical recycling processes. However, they require mitigation of unique pollutants arising from the decomposition of solid-state electrolytes, such as sulfur oxides and rare earth complexes.

Due to the complex composition of solid-state batteries, the recovery of valuable metals from spent solid-state batteries via hydrometallurgical processes—and their subsequent re-synthesis into cathode materials and solid-state electrolytes—remains technically challenging.[81] Sulfide-based solid electrolytes are air-sensitive and readily undergo hydrolysis in the presence of moisture, releasing toxic H_2S gas. Consequently, hydrometallurgical processes are unsuitable for the recycling of sulfide-based solid-state batteries. For polymer-based solid electrolytes, even minimal dissolution into aqueous solutions can significantly alter solution viscosity and interfere with the precipitation of metal ions—such as NCM—thereby complicating the recovery of valuable metals from spent solid-state batteries using this technology.[82] For oxide-based solid electrolytes, because the solubility product constants $\left(K_{sp}\right)$ of the metal hydroxides of constituent elements—such as La, Zr, and Ti—are lower than those of cathode metals—such as NCM—selective precipitation of these metal ions can be achieved by adjusting the pH of the leaching solution, the choice of extractant, and the precipitation conditions. For certain

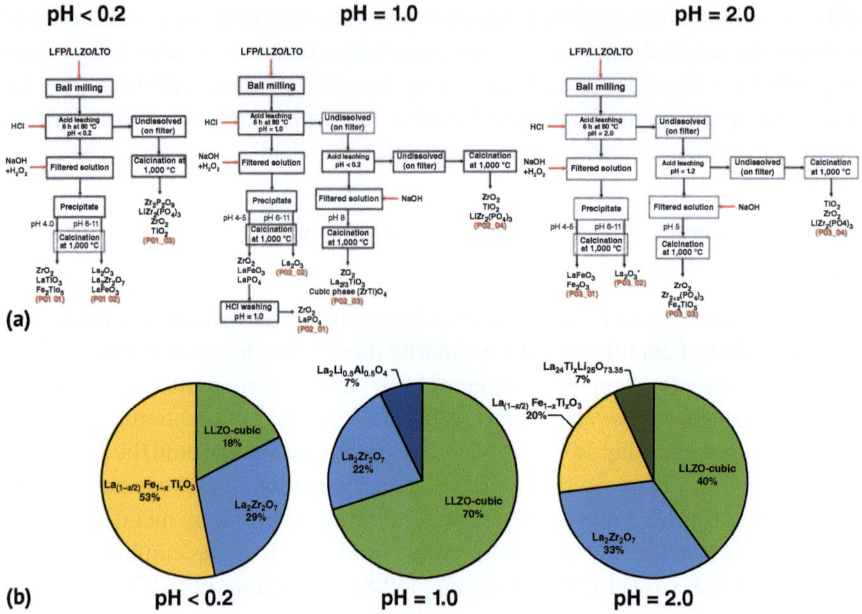

Figure 4.8 (a) Flow chart of the acid leaching/alkali precipitation process for the LFP/LLZO/LTO mixture using HCl (for leaching) with pH < 0.2, pH = 1.0, and pH = 2.0. (b) Relative weight fraction of the recycled LLZO as compared to other products at each leaching scenario (pH < 0.2, pH = 1.0, and pH = 2.0).

oxide-based solid electrolytes that are synthesized via coprecipitation, the solid electrolyte and cathode materials can be separately recovered through this process and subsequently re-synthesized by simple heat treatment and chemical re-lithiation.[83]

Recently, Friedrich et al. proposed a recycling strategy for oxide-based solid-state batteries comprising inert crushing, thermal treatment, sorting as a pretreatment step and multistep hydrometallurgical processes.[84] Subsequently, Nowroozi et al. employed HCl as the leaching agent and NaOH combined with H_2O_2 as the precipitating system to recover LLZO from the LFP/$Li_4Ti_5O_{12}$ composite via hydrometallurgical processing, as illustrated in Figure 4.8. They also found that variations in the concentration of the leaching solution significantly affect the purity of the final recovered product.[85] Although hydrometallurgical processes for LIBs have advanced rapidly, research on their effectiveness in recycling solid-state batteries remains limited. Therefore, developing tailored hydrometallurgical technologies for the recycling of spent solid-state batteries prior to their large-scale obsolescence is of critical importance.

4.2.8 Analysis of Hydrometallurgical Recycling for Solid-state Batteries

The hydrometallurgical recycling process recovers valuable metals from solid-state batteries through chemical leaching and separation techniques, but its technical bottlenecks primarily stem from the high corrosivity of leaching agents, interference by emerging impurities, and a significant increase in energy consumption during pretreatment.

The chemical inertness and compositional specificity of solid electrolytes have substantially altered the reaction pathways of conventional hydrometallurgical processes. For instance, the dense crystalline structure of oxide-based solid electrolytes (e.g., LLZO) exhibits minimal reactivity toward conventional acid leaching systems. High-concentration hydrofluoric acid or mixed acidic solutions are required to disrupt their lattice structure, leading to higher reagent consumption compared to the recycling process for conventional LCO batteries. During this process, La^{3+} and Zr^{4+} in LLZO co-dissolve with target metals such as Co^{2+} and Li^{+}, forming stable complexes including LaF_3 and ZrF_6^{2-} that compete with cobalt for the active sites of the extractant. This necessitates additional multistage pH adjustments, prolonging the separation process and increasing the likelihood of lithium being co-precipitated into the residue phase by fluorides, thereby causing an imbalance in the recovery rates of lithium and cobalt. The hydrometallurgical processing of sulfide-based solid electrolytes carries the risk of releasing highly toxic gases. It rapidly decomposes into H_2S upon exposure to acid or water. Therefore, pre-oxidation in a closed reactor is required to convert sulfur species into SO_4^{2-}. However, excessive oxidants can lead to over-oxidation of the cathode material (e.g., $LiCoO_2$), transforming Co^{3+} into higher oxidation states and necessitating a secondary reduction step to adjust the redox potential of the leaching solution, thereby increasing reagent costs. Meanwhile, the high concentration of sulfate ions in the leaching solution interferes with the subsequent electrowinning process, necessitating the addition of a precipitation or ion exchange unit, thereby increasing equipment investment. The dissolution of polymer-based electrolytes relies on strong polar solvents, but these exhibit poor compatibility with PVDF binders. Residual $TFSI^-$ $(CF_3\ SO_2^-)$ anions generate HF and SO_2 during acid leaching, leading to reactor wall corrosion and necessitating the implementation of a fluoride removal system, thereby increasing sludge treatment costs.

In addition, the integrated architecture of solid-state batteries increases the complexity of pretreatment. The electrode and electrolyte layers are tightly bonded through hot pressing or sintering. Conventional crushing methods are ineffective in achieving their efficient separation, necessitating the introduction of high-pressure roller milling or deep cryogenic crushing using liquid nitrogen, thereby increasing energy consumption relative to conventional recycling processes. The nanoscale rare earth dust generated during mechanical activation tends to adsorb onto the surface of cathode particles, forming an "inert shell layer" that impedes acid solution penetration. Consequently, microwave-assisted oxidation must be coupled to disrupt the interfacial barrier, although the investment in microwave equipment remains relatively high. Ultimately, the hydrometallurgical process for recycling solid-state batteries faces three major challenges: high reagent toxicity, prolonged separation procedures, and doubled energy consumption in pretreatment. There is an urgent need to develop green leaching agents, selective separation technologies, and intelligent pretreatment systems to overcome the industrialization bottleneck.

4.2.9 Demand for Hydrometallurgical Recycling for Cathode Materials from Spent LIBs

Hydrometallurgical technology, primarily based on leaching, typically dissolves all valuable metals simultaneously, resulting in a complex solution composition with impurities such as Al, Fe, Cu, Ca, Mg, F, and S. Al and Cu primarily originate from fragments of the cathode current collector, battery casing, and anode current collector. The Fe impurities in the leachate originate from the mechanical components of the recycling equipment and from the battery packaging materials. These impurities consume chemical reagents, cause precious metal losses, and exhibit low recycling value, reducing the purity of the recycled products. In addition, metal impurities in the cathode material dissolve at high potentials and subsequently deposit on the anode as metallic particles at low potentials, posing a risk of internal short circuits in LIBs. On the other hand, several studies have demonstrated that trace amounts of Al, Mg, and Fe can serve as dopants to enhance the electrochemical performance of active materials, whereas excessive impurities are detrimental to LIBs.[11] The tolerance level of impurities presents a challenging control issue. Therefore, impurity control in hydrometallurgical processes is a critical factor influencing the value of subsequent recycled products.

Although numerous studies have focused on separating these impurities from spent cathode materials during the pretreatment stage, subsequent research has revealed that the impurity levels remain low at this stage. Nevertheless, separation and purification processes are still necessary to further reduce impurity content. For metallic impurities such as Fe, Al, and Cu, the chemical precipitation method is initially employed based on the differences in solubility of their compounds at varying pH values. If the subsequent step involves a resynthesis process, solvent extraction is required to achieve deeper impurity removal due to the regenerated material's high sensitivity to impurities. In addition, to prevent the co-dissolution of various metals, stepwise leaching and selective leaching are widely employed as leaching techniques. Liu et al. proposed a process involving reduction and calcination of NCM to convert lithium into water-soluble Li_2CO_3 while transforming NCM into water-insoluble compounds. Subsequently, Li was recovered through water leaching, and NCM was extracted via acid leaching. In ammonia-based leaching systems, selective leaching is commonly achieved due to the preferential chelation between ammonium ions and transition metal ions.[12] Ku et al. developed an ammonia leaching system using ammonia as the leaching agent, $(NH_4)_2SO_3$ as the reducing agent, and $(NH_4)_2CO_3$ as the pH buffer. The leaching behavior of Ni, Co, Mn, Al, and Cu from spent cathode materials was investigated. Under optimal leaching conditions, this method achieves complete dissolution of Cu and Co, partial extraction of Ni and minimal leaching of Mn and Al.

In the hydrometallurgical recycling process of spent LIBs, the demand for separation and recovery technologies is shifting toward higher efficiency, lower cost, and greater environmental friendliness. To improve the recycling efficiency and purity of recovered metals, future technological developments are primarily focused on the following aspects:

First, green and environmentally friendly recycling technologies will become a key requirement. As environmental regulations grow increasingly stringent, the chemical reagents and solvents used in traditional hydrometallurgical recycling processes pose significant challenges in terms of environmental pollution and wastewater management. To reduce the environmental burden, future separation and recycling technologies should prioritize the development of green solvents and chemical reagents that are low in toxicity and readily recyclable. These solvents should not only exhibit high selectivity but also minimize waste liquid generation, ensuring a more environmentally friendly recycling process. In addition, efficient wastewater and waste gas treatment technologies represent essential future development directions to ensure that by-products generated during the recycling process are safely managed and resources are effectively recovered.

Second, enhancing the efficiency and selectivity of metal separation is central to technological advancement. To improve the purity of recovered metals and minimize resource waste, future technologies must focus on optimizing separation performance. For instance, more efficient sedimentation and solvent extraction methods should be developed to improve the separation efficiency of metal ions. Meanwhile, future technological requirements will also encompass the integration of multiple separation techniques into combined processes, enabling more precise metal separation and enhancing both recovery rates and metal purity. For membrane separation technology, developing fouling-resistant and corrosion-resistant membrane materials with improved cost-effectiveness will be a key future direction, thereby reducing equipment and maintenance costs in the overall recycling process. Reducing costs is crucial to enabling the industrialization of spent LIB recycling. To improve the economic viability of the recycling process, future technological developments will focus on reducing the costs of separation materials and equipment. For instance, developing cost-effective ion exchange resins and membrane materials with extended service life can reduce replacement frequency and operational expenses. Optimizing existing processes, simplifying operational procedures, and minimizing chemical and energy consumption will also significantly reduce overall recycling costs. In addition, driven by technological advancements and growing market demand, future recycling technologies should incorporate more flexible process designs to meet the diverse recycling requirements of different types of spent LIBs.

Finally, promoting the establishment of a closed-loop recycling system for separation and recovery is essential to future technological development. In hydrometallurgical recovery processes, the waste liquid typically contains valuable metal ions that have not been fully recovered, such as lithium, cobalt, and nickel. Efficiently recovering these metal ions is crucial to enhancing resource utilization. Therefore, the development of efficient secondary recovery technologies—particularly for residual metals in leaching solutions—is essential. Through selective adsorption, solvent extraction, and ion exchange, metal recovery rates can be improved while resource waste is minimized. Establishing a closed-loop recycling system not only maximizes the recovery of metals from waste streams but also enhances overall recycling efficiency, thereby promoting resource circularity and sustainable development.

4.3 Metallurgical Recycling Development Trends Across Different Battery Systems

4.3.1 LFP Batteries

LFP batteries have gained a significant position in the fields of new energy vehicles and energy storage due to their low cost and high safety, but the economic viability of recycling them remains a key bottleneck for the industry. Currently, both mainstream pyrometallurgical and hydrometallurgical recycling processes exhibit significant limitations. Pyrometallurgical recycling recovers metals through high-temperature melting, offering advantages such as a simple process flow and high processing efficiency. However, in the treatment of LFP batteries, lithium is highly volatile under elevated temperatures, while iron and phosphorus tend to form Fe-P alloys with low economic value, resulting in overall revenues that are insufficient to cover recycling costs. Although the hydrometallurgical method can extract lithium, iron, and phosphorus from LFP batteries through acid leaching, it faces challenges such as significant interference from impurities and complex separation processes. For instance, the precursor of $FePO_4$ requires repeated purification, leading to high consumption of acid and alkali reagents. Moreover, pretreatment of LFP batteries constitutes a substantial portion of the overall recycling cost. From the perspective of scientific researchers, the focus of future LFP battery recycling technology lies in innovating pyrometallurgical and hydrometallurgical processes, overcoming existing technical bottlenecks, and improving resource recycling efficiency. The core challenges of pyrometallurgical processes are lithium volatilization losses and the low economic value of iron-phosphorus slag. Current process development primarily focuses on the following three aspects: (1) Develop targeted solidification technologies to suppress lithium volatilization: by developing novel smelting additives—such as calcium-based or silicon-based solidifiers—stable compounds (e.g., Li_2CaO_3 and Li_4SiO_4) can be formed with lithium at high temperatures, preventing its release into flue gas. This approach aims to increase the lithium recovery rate from currently less than 30% to over 60%. (2) Optimize the process to lower the smelting temperature: by integrating microwave or plasma-assisted heating technologies, the pyrometallurgical reaction temperature can be reduced to below 1,100 °C—compared to the conventional process requiring approximately 1,500 °C—thereby reducing energy consumption and mitigating lithium volatilization. (3) Multistage smelting and coordinated lithium recovery: by integrating pyrometallurgical and hydrometallurgical processes, nonvolatile components such as iron and phosphorus are first recovered through high-temperature melting, followed by low-temperature hydrometallurgical leaching of the residue for secondary lithium extraction, enabling stepwise utilization of all valuable components. The key challenges in hydrometallurgical processes lie in environmental pollution associated with strong acid reagents and the high cost of purifying and separating LFP. Future research and development priorities include: (1) Development of green leaching systems: utilizing low-toxicity organic acids (e.g., citric acid and oxalic acid) or bioleaching agents (such as metabolic products of iron-oxidizing bacteria) to replace sulfuric or hydrochloric acid, reducing both the acidity of the leaching solution and the risk of heavy metal pollution. (2) Selective lithium extraction technology: develop high-affinity adsorption materials (e.g., titanium-based lithium-ion sieves) or membrane-based separation systems to enable selective recovery of lithium ions directly from the leachate, avoiding separation challenges arising from the co-leaching of iron and phosphorus. (3) Short-process regeneration technology of LFP: based on electrochemical redox reactions or solid-state ion transport mechanisms, waste LFP can be directly reconstructed into high-purity LFP cathode material, bypassing the hydrometallurgical purification step for the LFP intermediate and significantly reducing recycling costs.

Starting from industrial demands for commercialization and implementation, future iterations of LFP battery recycling technology will focus on cost optimization in pyrometallurgical processes and process simplification in hydrometallurgical processes to address the stability and economic challenges of existing production lines. The pyrometallurgical process faces the following industrialization challenges: (1) Integrated development of the lithium recovery system from flue gas: develop waste gas treatment equipment—such as multistage condensation towers and lithium salt capture filters—that is compatible with high-temperature smelting furnaces to enable large-scale recovery of volatile lithium compounds (e.g., Li_2CO_3 and Li_3PO_4) and integrate lithium enrichment and purification modules to reduce lithium recovery costs. (2) High-value conversion process of iron-phosphorus slag: upgrading traditional Fe-P alloys into functional materials—such as Fe-P-based catalysts for wastewater treatment or raw materials for LFP precursor synthesis—significantly enhances their added value. (3) Development of fully automatic high-temperature smelting equipment: by optimizing refractory materials and implementing intelligent temperature control technology, the service life of the smelting furnace is extended, and unit energy consumption is reduced. The future development trends of hydrometallurgical processes focus on two key aspects: process intensification and resource closure. (1) Integrated leaching/regeneration equipment: develop a continuous production line that integrates acid leaching, filtration, and electrochemical lithium extraction. By intelligently coordinating process parameters—such as pH, temperature, and real-time potential regulation—the leaching time can be significantly reduced. (2) Impurity-targeted removal technology: introducing ion-imprinted polymers (IIPs) or selective complexing extractants to preferentially adsorb interfering impurities such as copper and aluminum from the leaching solution, reducing reagent consumption in subsequent separation processes. (3) Regionalized material circulation network: recycled LFP is directly supplied to nearby battery manufacturing facilities, while the acid solution achieves over 90% reuse through membrane distillation, establishing a local closed-loop cycle of "recycling–regeneration–production" and reducing both logistics and environmental compliance costs.

4.3.2 Ternary LIB

NCM batteries have dominated the premium electric vehicle market due to their high energy density, but the coexistence of multiple metals—such as NCM and lithium—renders the recycling process significantly more complex. Although current pyrometallurgical recovery enables the extraction of Ni-Co alloys via high-temperature reductive smelting, lithium—due to its low boiling point—volatilizes and reports to the slag primarily as lithium oxide, necessitating a secondary hydrometallurgical leaching step that doubles both energy consumption and processing costs. Hydrometallurgical recovery employs acidic reagents for leaching, but the leaching efficiencies of nickel and cobalt vary significantly, and acid as well as alkali consumption is high. Moreover, treating and recovering heavy metals and salts from wastewater entails substantial costs. Additionally, residual aluminum impurities in the feedstock can lead to aluminum ion contamination, requiring the addition of fluoride for precipitation and impurity removal, increasing overall recovery costs.

From the perspective of scientific researchers, future breakthroughs in NCM battery recycling technology will focus on innovations in the mechanisms of pyrometallurgical and hydrometallurgical processes, aiming to overcome challenges related to metal component separation, high energy consumption, and secondary pollution. The core challenges of pyrometallurgical processes lie in the high-temperature oxidation losses of NCM, as well as fluoride and phosphorus emissions in flue gas. Key research directions should therefore focus on: (1) Develop low-temperature and oxygen-free smelting technology: develop plasma smelting or microwave pyrolysis processes under inert gas protection (reaction temperature $< 1{,}000\,°C$) to suppress the oxidation and volatilization of nickel and cobalt in ternary cathode materials such as NCM622, and enable directional enrichment of metal alloy phases with Ni-Co-Mn alloy purity exceeding 95%. (2) Develop slag component regulation technology: by adding fluxing agents—such as Al_2O_3-SiO_2-based glassy materials—to complex residual fluorine and lithium impurities from the electrolyte, a stable glassy phase is formed to immobilize pollutants while enhancing metallic phase separation efficiency. (3) Innovate lithium capture and recovery technology through volatilization: in response to lithium's high-temperature volatility, design an aerosol condensation–chemical adsorption system to recover Li_2CO_3 or LiF compounds from flue gas, significantly increasing the lithium recovery efficiency in pyrometallurgical processes. The hydrometallurgical process needs to overcome the bottlenecks of the difficulty in regenerating acidic reagent leaching solutions and the high complexity of multi-metal separation. The future development priorities include: (1) Low-eutectic solvent (DES) leaching system: a novel green deep eutectic solvent system is employed to replace conventional sulfuric or hydrochloric acid, enabling efficient and selective leaching of nickel and cobalt via hydrogen bonding while reducing the generation of hazardous liquid waste. (2) Combined precipitation and electrochemical separation technology: develop amino carboxylic acid-based chelating agents to selectively bind cobalt and nickel ions, and directly synthesize metallic or oxide powders via a three-electrode electrodeposition system, bypassing the extraction step and reducing organic reagent consumption. (3) Short-process closed-loop regeneration technology: based on molten salt electrolysis or electrochemical redox mechanisms, spent ternary cathode materials are directly reduced to precursors (e.g., $Ni_{0.6}Co_{0.2}Mn_{0.2}(OH)_2$), enabling in situ regeneration of cathode materials and reducing energy consumption associated with hydrometallurgical purification.

Starting from industrial practical demands, the future development of NCM battery recycling technology will focus on enhancing the scalability and economic viability of hydrometallurgical processes, aiming to reduce operational costs of existing production lines and overcome current limitations in recycling efficiency. Key areas for technological upgrading include: (1) Integrated smelting and separation system: develop multistage melting furnaces—such as integrated shaft furnace and electric arc furnace systems—to achieve pre-separation of aluminum and copper foils from ternary black mass through gradient temperature control, while concentrating NCM into Fe-based alloy phases suitable for direct use as raw materials in stainless steel or specialty alloys, reducing purification costs. (2) Develop equipment for flue gas purification and lithium recovery: integrate a high-temperature electrostatic precipitator with an alkaline liquid spray tower to simultaneously capture HF gas and lithium compounds and purify the lithium-rich waste stream into battery-grade lithium carbonate. (3) Optimize the carbothermal reduction process: utilize waste graphite from anode materials as a reducing agent to reduce high-valent metal oxides during the pyrometallurgical stage, decreasing coke consumption and lowering carbon emission intensity. The development path of hydrometallurgical processes focuses on: (1) Developing a coordinated leaching and separation process for complex metal systems: designing an integrated leaching-extraction-electrolysis system to enhance acid utilization through countercurrent leaching and achieving one-step separation of NCM using multistage centrifugal contactors. (2) Develop electrochemical regeneration and recycling technology for acidic reagents: implement a bipolar membrane electrodialysis system to decompose and regenerate sulfuric acid ($pH < 1$) from leaching waste streams into H^+ and SO_4^{2-} for reuse, reducing consumption of fresh acid and enabling simultaneous recovery of residual lithium. (3) Develop a digital process monitoring platform: utilize AI algorithms to optimize leaching time, temperature and stirring rate in real time, reduce fluctuations in component leaching kinetics, and extend the continuous and stable operation cycle of the production line.

4.3.3 LCO Batteries

LCO batteries are primarily used in the 3C consumer electronics sector. Due to their high cobalt content (approximately 60%), they possess significant recycling value. However, environmental concerns associated with their recycling processes are prominent. Pyrometallurgical recycling recovers cobalt through reductive smelting. However, the process generates highly toxic slag containing cobalt and copper, with concentrations of arsenic, lead, and other impurities exceeding regulatory limits by up to 50 times, and the disposal cost accounts for 25% of the total recycling cost.

Hydrometallurgical recycling employs acidic reagents to leach cobalt; however, this process simultaneously causes the dissolution of aluminum foil, resulting in an Al^{3+} concentration in the solution reaching up to 2,000 ppm, necessitating the addition of supplementary reagents for its removal. In addition, when LCO batteries are mixed with other battery types, the resulting leachate contains excessive iron and manganese impurities, requiring removal through multiple purification stages and significantly increasing recycling costs.

From the perspective of scientific researchers, future innovations in LCO battery recycling technology will focus on optimizing the efficiency of pyrometallurgical and hydrometallurgical processes. The key challenges lie in improving cobalt efficiency, minimizing lithium loss, and controlling environmental pollution. The challenge of pyrometallurgical processes lies in the loss of cobalt reducibility at high temperatures and the emission of chlorides and fluorides. The primary focus of scientific research is on: (1) Dynamic regulation technology for reducing atmosphere: develop a low-temperature (800–1,000 °C) reduction smelting process under the protective gas atmosphere. By precisely controlling the oxygen partial pressure, the oxidation of cobalt into refractory CoO or Co_3O_4 is suppressed, enabling metallic cobalt to be enriched in elemental or alloyed forms (e.g., Co–Fe phase), significantly enhancing the direct recovery efficiency of cobalt. (2) Volatile lithium directional enrichment: introduce alkali metal capture agents into the smelting system to react with high-temperature gaseous lithium compounds (e.g., Li_2O and LiCl), forming stable silicates such as $LiAlSiO_4$, and enable secondary lithium recovery through subsequent hydrometallurgical leaching, ultimately enhancing the overall lithium recovery efficiency. (3) Regulate the mineral phase of toxic components in slag: utilize borax or sodium carbonate additives to immobilize fluorine and phosphorus impurities generated during the thermal decomposition of the electrolyte into glassy silicates (e.g., CaF_2–SiO_2 solid solution), preventing the formation of dioxins and other hazardous pollutants, enhancing slag stability and reducing its leaching toxicity.

From the perspective of commercialization and industrial implementation, LCO battery recycling technology will focus on overcoming the challenges of large-scale continuous operation in pyrometallurgical processes and the economic viability of hydrometallurgical processes, promoting cost reduction and efficiency improvement in industrialization. Future advancements in pyrometallurgy will primarily focus on: (1) Continuous extraction technology for molten cobalt metal: construct an integrated system comprising a rotary kiln, electric arc furnace, and lithium capture tower to enable full-component stepwise recovery through staged pyrolysis and achieve efficient purification and separation of electrode materials and current collectors, improved cobalt alloy purity, and enhanced single-line processing capacity. (2) High-temperature flue gas resource utilization process: deploy a three-stage treatment system comprising a quench tower, bag filter, and lithium absorption tower to condense LiF and lithium vapor from smelting flue gas into lithium-rich powder. After carbonation conversion, battery-grade lithium carbonate is produced, reducing the cost of flue gas treatment. (3) Low-carbon energy composite technology: by employing a plasma smelting system powered by green electricity and utilizing biomass-derived charcoal as a substitute for coke as the reducing agent, carbon emissions from the pyrometallurgical recycling of cobalt-containing LCO batteries are significantly reduced. The technological breakthroughs in hydrometallurgical processes will focus on: (1) Targeted extraction of cobalt and lithium: develop customized chelating resins to selectively adsorb cobalt ions via pH gradient control, followed by desorption using low-concentration sulfuric acid to obtain a high-purity cobalt solution; (2) Electrochemical/membrane separation hybrid process: a dual-membrane electrolytic cell system comprising an anion exchange membrane and a nanofiltration membrane is employed. In the anode compartment, LCO is directly electro-oxidized to release Co^{3+} while simultaneously regenerating the electrolyte. In the cathode compartment, lithium ions are retained by the nanofiltration membrane to produce LiOH solution (lithium purity $\geq$ 99%), reducing overall acid consumption in the process. (3) AI-driven leaching kinetics optimization: by applying machine learning models to enable real-time analysis of LCO particle size, leaching agent concentration, and reaction interface parameters, the stirring rate and temperature are dynamically adjusted to reduce fluctuations in cobalt leaching efficiency and shorten the processing time per leaching batch.

4.3.4 Solid-state Battery

Solid-state batteries, offering high energy density and enhanced safety, are increasingly being adopted in electric vehicles and high-end energy storage applications. However, their recycling poses significant challenges, including high energy consumption, complex separation of components, and the management of emerging pollutants. The high thermal stability (melting point of oxide electrolyte > 1,600 °C) and chemical inertness of solid electrolytes significantly increase the energy demand of pyrometallurgical smelting, while the hydrometallurgical process must address the release of highly toxic H_2S from sulfide decomposition and the challenges associated with separating rare earth elements such as lanthanum and zirconium. Future technological innovations will concentrate on developing low-temperature fluxes, establishing targeted leaching systems and optimizing direct regeneration processes, advancing the transformation of solid-state battery recycling toward higher efficiency and reduced carbon emissions. From a scientific research perspective, recycling of solid-state batteries requires overcoming the inherent limitations of pyrometallurgical and hydrometallurgical processes, with primary emphasis on efficient metal extraction, effective suppression of pollutants, and process streamlining. (1) Low-temperature decomposition technology for solid-state electrolytes: develop composite fluxes to lower pyrometallurgical smelting temperatures and simultaneously introduce a mixed reducing atmosphere to prevent the formation of refractory phases between rare earth oxides and cobalt or nickel, enhancing cobalt recovery efficiencies. (2) Directional transformation of sulfides and targeted lithium recovery: design a photocatalytic oxidation–electrochemical coupling system to minimize H_2S generation while simultaneously enriching lithium ions via nanofiltration membranes in the cathode region, thereby enhancing lithium purity. (3) Harmless treatment of polymer electrolytes: develop a supercritical water oxidation process to decompose PEO-LiTFSI into CO_2, H_2O, and recyclable LiF, enhance the fluorine fixation efficiency and simultaneously degrade the PVDF binder into harmless short-chain hydrocarbons, avoiding the dioxin formation risk associated with traditional incineration.

From an industrialization perspective, enterprises must address the challenges of large-scale continuous production, cost control, and environmental compliance by focusing on key technological and operational advancements in the following areas: (1) Pyrometallurgical-hydrometallurgical coupling production line integration: construct an integrated equipment of "rotary kiln pre-sintering + electric arc furnace smelting + pressure leaching," to process oxide-based solid-state batteries in a graded manner. In the pre-sintering stage, the electrolyte-electrode interface is broken down. In the smelting stage, high-purity Co-Ni alloys are produced. In the hydrometallurgical stage, rare earths (La and Zr) and lithium are recovered, enhancing the single-line processing capacity. (2) Waste gas and residue resource utilization system: implement a three-stage purification system comprising a quench tower, catalytic oxidation unit, and lithium absorption tower to convert H_2S in flue gas from sulfide smelting into elemental sulfur, improving sulfur recovery efficiency. Meanwhile, the condensed lithium-rich dust is converted into battery-grade Li_2CO_3 via carbonation, enhancing the value-added potential of by-products. (3) Intelligent hydrometallurgical process optimization: AI models are employed for real-time control of leaching parameters—such as acid concentration, redox potential, and temperature—and enable dynamic adjustment of oxidant dosage in sulfide-based battery (LGPS) processing to suppress cobalt over-oxidation, reducing fluctuations in cobalt leaching efficiency and minimizing acid consumption.

4.4 Chapter Summary

4.4.1 Demand for Pyrometallurgical Recycling Technology

Pyrometallurgy must address the core challenge of low metal recovery efficiencies and high energy consumption. In the current process, lithium loss due to high-temperature volatilization is significant, with recovery efficiencies generally below 50%; manganese is difficult to recover because it forms stable oxides; and although nickel and cobalt can be enriched in alloy form, this does not compensate for the overall resource waste. Take ternary batteries as an example; after smelting, a significant portion of lithium reports to the slag. In LFP batteries, the Fe-P alloy has low added value (3,000–4,000 CNY/ton), and lithium reacts with silicon and aluminum components in the slag to form insoluble compounds such as Li_2SiO_3, further reducing recovery efficiency. Thus, it is urgently necessary to develop a new generation of efficient smelting systems: (1) Develop molten salt-assisted, low-temperature smelting technology by combining low-melting-point additives (such as carbonates) with carbon-based reducing agents to suppress lithium volatilization and reduce the smelting temperature below 1,100 °C. (2) Optimize the slag composition by introducing rare earth metal oxides and other fluxes to enhance manganese reducibility and reduce its retention in the slag phase. (3) Innovate the separation process by integrating electromagnetic stirring or rotary furnaces to enhance metal-slag separation efficiency, and simultaneously develop solid lithium capture agents for deep lithium extraction from the slag. In addition, a coordinated smelting system for spent batteries and primary ores should be established. By matching the Fe/S and CaO/SiO_2 ratios of copper-nickel-cobalt ores with those of battery materials, an integrated multisource processing route—such as top-blown bath smelting—can be designed. This approach not only reduces energy consumption but also enhances the multi-metal migration mechanism, enabling targeted lithium transfer and simultaneous control of harmful gases (e.g., SO_2 and HF).

4.4.2 Demand for Hydrometallurgical Recycling Technology

Hydrometallurgy must overcome three major limitations: limited lithium recovery, impurity interference, and insufficient product purity. Lithium leaching is hindered by the material structure (e.g., Mn^{3+} disproportionation blocking diffusion channels in LMO), and divalent manganese in the leaching solution competes with lithium for precipitation, which, together with the adsorption of carbonaceous materials from the anode, contributes to reduced recovery efficiency. Meanwhile, targeted impurity removal remains insufficient—the pH ranges for precipitation of iron, aluminum, copper, and other metal hydroxides overlap with those of nickel and cobalt, resulting in a 5–10% loss of target metals, while phosphate complex colloids in the LFP system further increase separation difficulty. To enhance materialization value, breakthroughs are required across multiple levels. (1) Develop dynamic pH control strategies and highly selective adsorbents (e.g., functionalized MOFs) to enable preferential lithium release and precise capture. (2) Develop targeted impurity removal technology by designing specific complexing agents (e.g., thiourea for copper capture and EDTA for iron fixation), Fe^{3+}-imprinted IIPs with an adsorption capacity of 120 mg/g, and integrating nanofiltration-AI systems for dynamic removal of trace impurities. (3) By developing intelligent crystallization and integrated regeneration processes, battery-grade NCM precursor with a purity exceeding 99.5% can be directly synthesized through coupling temperature control with nucleation factors, thereby avoiding the cost increase and performance degradation associated with multistage purification, and enabling the closed-loop transformation of hydrometallurgical process products into high-value materials.

References

1 Z. Zeng, P. Xu, J. Li, C. Yi, W. Zhao, W. Sun, X. Ji, Y. Yang and P. Ge, *Advanced Functional Materials*, 2024, **34**, 2308671.

2 R.-C. Wang, Y.-C. Lin and S.-H. Wu, *Hydrometallurgy*, 2009, **99**, 194–201.

3 T. Georgi-Maschler, B. Friedrich, R. Weyhe, H. Heegn and M. Rutz, *Journal of Power Sources*, 2012, **207**, 173–182.

4 X. Li, F. Zhou, S. Gao, J. Zhao, D. Wang and H. Yin, *Waste Management*, 2022, **153**, 347–354.

5 J. Yan, H. Huang, J. Tong, W. Li, X. Liu, H. Zhang, H. Huang and W. Zhou, *Interdisciplinary Materials*, 2022, **1**, 330–353.

6 S. Maroufi, M. Assefi, R. Khayyam Nekouei and V. Sahajwalla, *Sustainable Materials and Technologies*, 2020, **23**, e00139.

7 J. Xiao, J. Li and Z. Xu, *Environmental Science & Technology*, 2017, **51**, 11960–11966.

8 Y. Yue, S. Wei, B. Yongjie, Z. Chenyang, S. Shaole and H. Yuehua, *ACS Sustainable Chemistry & Engineering*, 2018, **6**, 10445–10453.

9 T. Nshizirungu, M. Rana, M. I. H. Khan, Y. T. Jo, S.-J. Park and J.-H. Park, *Journal of Environmental Chemical Engineering*, 2023, **11**, 109160.

10 Y. Zhang, W. Wang, Q. Fang and S. Xu, *Waste Management*, 2020, **102**, 847–855.

11 Q. Peng, X. Zhu, J. Li, Q. Liao, Y. Lai, L. Zhang, Q. Fu and X. Zhu, *Waste Management*, 2021, **134**, 100–109.

12 H. Chen, P. Hu, D. Wang and Z. Liu, *Hydrometallurgy*, 2022, **210**, 105865.

13 H. Park, O. Tamwattana, J. Kim, S. Buakeaw, R. Hongtong, B. Kim, P. Khomein, G. Liu, N. Meethong and K. Kang, *Advanced Energy Materials*, 2020, **11**, 2003039.

14 Y. Yang, X. Meng, H. Cao, X. Lin, C. Liu, Y. Sun, Y. Zhang and Z. Sun, *Green Chemistry*, 2018, **20**, 3121–3133.

15 H. Li, S. Xing, Y. Liu, F. Li, H. Guo and G. Kuang, *ACS Sustainable Chemistry & Engineering*, 2017, **5**, 8017–8024.

16 J. Zhang, J. Hu, Y. Liu, Q. Jing, C. Yang, Y. Chen and C. Wang, *ACS Sustainable Chemistry & Engineering*, 2019, **7**, 5626–5631.

17 H. Jin, J. Zhang, C. Yang, L. Ma, Y. Chen and C. Wang, *Chemical Engineering Journal*, 2023, **460**, 141805.

18 Z. Li, D. Liu, J. Xiong, L. He, Z. Zhao and D. Wang, *Waste Management*, 2020, **107**, 1–8.

19 R. C. Massé, C. Liu, Y. Li, L. Mai and G. Cao, *National Science Review*, 2017, **4**, 26.

20 S. Tao, J. Li, L. Wang, L. Hu and H. Zhou, *Ionics*, 2019, **25**, 5643–5653.

21 E. Fan, L. Li, X. Zhang, Y. Bian, Q. Xue, J. Wu, F. Wu and R. Chen, *ACS Sustainable Chemistry & Engineering*, 2018, **6**, 11029–11035.

22 J. Kumar, X. Shen, B. Li, H. Liu and J. Zhao, *Waste Management*, 2020, **113**, 32–40.

23 H. Shentu, B. Xiang, Y.-J. Cheng, T. Dong, J. Gao and Y. Xia, *Environmental Technology & Innovation*, 2021, **23**, 101569.

24 T. Yan, S. Zhong, M. Zhou, X. Guo, J. Hu, F. Wang, F. Zeng and S. Zuo, *Nanotechnology Reviews*, 2020, **9**, 1586–1593.

25 D.-y. Wu, D.-x. Wang, Z.-q. Liu, S. Rao and K.-f. Zhang, *Transactions of Nonferrous Metals Society of China*, 2022, **32**, 2071–2079.

26 X. Zhou, W. Yang, X. Liu, J. Tang, F. Su, Z. Li, J. Yang and Y. Ma, *Waste Management*, 2023, **155**, 53–64.

27 J. Yu, X. Wang, M. Zhou and Q. Wang, *Energy & Environmental Science*, 2019, **12**, 2672–2677.

28 B. Zhang, X. Qu, J. Qu, X. Chen, H. Xie, P. Xing, D. Wang and H. Yin, *Green Chemistry*, 2020, **22**, 8633–8641.

29 L. Li, J. Lu, L. Zhai, X. Zhang, L. Curtiss, Y. Jin, F. Wu, R. Chen and K. Amine, *CSEE Journal of Power and Energy Systems*, 2018, **4**, 219–225.

30 K. Ji, J. Han, A. Hirata, T. Fujita, Y. Shen, S. Ning, P. Liu, H. Kashani, Y. Tian and Y. Ito, *Nature Communications*, 2019, **10**, 275.

31 S. Sinha, S. S. Nene, M. Frank, K. Liu, P. Agrawal and R. S. Mishra, *Scientific Reports*, 2019, **9**, 13185.

32 L. Yang, Y. Feng, C. Wang, D. Fang, G. Yi, Z. Gao, P. Shao, C. Liu, X. Luo and S. Luo, *Chemical Engineering Journal*, 2022, **431**, 133232.

33 X. J. Wang, S. L. Zheng, Y. Zhang, Y. Zhang, S. Qiao, Z. Q. Long, B. Zhao and Z. F. Li, *Waste Management*, 2022, **153**, 31–40.

34 J. Zhou, C. Xing, J. Huang, Y. Zhang, G. Li, L. Chen, S. Tao, Z. Yang, G. Wang and L. Fei, *Advanced Energy Materials*, 2023, **14**, 2302761.

35 G. Hu, K. Huang, K. Du, Z. Peng and Y. Cao, *Journal of Cleaner Production*, 2024, **434**, 140091.

36 V. A. Nikitina, S. Y. Vassiliev and K. J. Stevenson, *Advanced Energy Materials*, 2020, **10**, 1903933.

37 B. Swain, J. Jeong, J.-c. Lee, G.-H. Lee and J.-S. Sohn, *Journal of Power Sources*, 2007, **167**, 536–544.

38 L. Sun and K. Qiu, *Journal of Hazardous Materials*, 2011, **194**, 378–384.

39 D. A. Ferreira, L. M. Z. Prados, D. Majuste and M. B. Mansur, *Journal of Power Sources*, 2009, **187**, 238–246.

40 J. Wang, M. Chen, H. Chen, T. Luo and Z. Xu, *Procedia Environmental Sciences*, 2012, **16**, 443–450.

41 Z. Takacova, T. Havlik, F. Kukurugya and D. Orac, *Hydrometallurgy*, 2016, **163**, 9–17.

42 M. A. H. Shuva and A. S. W. Kurny, *Journal of The Institution of Engineers (India): Series D*, 2013, **94**, 13–16.

43 L. Xue, C. Wang, H. Liu, H. Li, T. Chen, Z. Shi, C. Qiu, M. Sun, Y. Huang and J. Huang, *Advanced Science*, 2022, **9**, 2202194.

44 M. K. Jha, N. Van Nguyen, J. C. Lee, J. Jeong and J. M. Yoo, *Journal of Hazardous Materials*, 2009, **164**, 948–953.

45 X. Chen, H. Ma, C. Luo and T. Zhou, *Journal of Hazardous Materials*, 2017, **326**, 77–86.

46 Y. Zheng, W. Song, W. T. Mo, L. Zhou and J. W. Liu, *RSC Advances*, 2018, **8**, 8990–8998.

47 R. Golmohammadzadeh, F. Rashchi and E. Vahidi, *Waste Management*, 2017, **64**, 244–254.

48 P. Meshram, B. D. Pandey and T. R. Mankhand, *Waste Management*, 2015, **45**, 306–313.

49 L. P. He, S. Y. Sun, X. F. Song and J. G. Yu, *Waste Management*, 2017, **64**, 171–181.

50 P. Meshram, Abhilash, B. D. Pandey, T. R. Mankhand and H. Deveci, *JOM*, 2016, **68**, 2613–2623.

51 Y. Yang, S. Xu and Y. He, *Waste Management*, 2017, **64**, 219–227.

52 P. Meshram, B. D. Pandey and T. R. Mankhand, *Chemical Engineering Journal*, 2015, **281**, 418–427.

53 W. Lv, Z. Wang, H. Cao, X. Zheng, W. Jin, Y. Zhang and Z. Sun, *Waste Management*, 2018, **79**, 545–553.

54 S. Yang, Q. Gao, Y. Li, H. Cai, X. Li, G. Sun, S. Zhuang, Y. Tong, H. Luo and M. Lu, *Journal of Energy Chemistry*, 2024, **93**, 24–31.

55 L. Yao, Y. Feng and G. Xi, *RSC Advances*, 2015, **5**, 44107–44114.

56 L. Li, Y. Bian, X. Zhang, Q. Xue, E. Fan, F. Wu and R. Chen, *Journal of Power Sources*, 2018, **377**, 70–79.

57 X. Chen, D. Kang, J. Li, T. Zhou and H. Ma, *Journal of Hazardous Materials*, 2020, **389**, 121887.

58 C. Wang, S. Wang, F. Yan, Z. Zhang, X. Shen and Z. Zhang, *Waste Management*, 2020, **114**, 253–262.

59 H.-x. Wei, L.-b. Tang, Y.-d. Huang, Z.-y. Wang, Y.-h. Luo, Z.-j. He, C. Yan, J. Mao, K.-h. Dai and J.-c. Zheng, *Materials Today*, 2021, **51**, 365–392.

60 Y. Chen, N. Liu, F. Hu, L. Ye, Y. Xi and S. Yang, *Waste Management*, 2018, **75**, 469–476.

61 H. Zou, E. Gratz, D. Apelian and Y. Wang, *Green Chemistry*, 2013, **15**, 1183.

62 Y. Liu, W. Chen, C. Yang, Q. Wei and M. Wei, *Journal of Power Sources*, 2018, **392**, 226.

63 W. Gao, J. Song, H. Cao, X. Lin, X. Zhang, X. Zheng, Y. Zhang and Z. Sun, *Journal of Cleaner Production*, 2018, **178**, 833–845.

64 J. Hu, J. Zhang, H. Li, Y. Chen and C. Wang, *Journal of Power Sources*, 2017, **351**, 192–199.

65 C. Yang, J. Zhang, B. Yu, H. Huang, Y. Chen and C. Wang, *Separation and Purification Technology*, 2021, **267**, 118609.

66 G. Zhang, X. Yuan, C. Y. Tay, Y. He, H. Wang and C. Duan, *Separation and Purification Technology*, 2023, **314**, 123555.

67 P. Liu, L. Xiao, Y. Tang, Y. Chen, L. Ye and Y. Zhu, *Journal of Thermal Analysis and Calorimetry*, 2018, **136**, 1323–1332.

68 P. Liu, L. Xiao, Y. Chen, Y. Tang, J. Wu and H. Chen, *Journal of Alloys and Compounds*, 2019, **783**, 743–752.

69 Y. Zhao, B. Liu, L. Zhang and S. Guo, *Journal of Hazardous Materials*, 2020, **384**, 121487.

70 F. Liu, C. Peng, Q. Ma, J. Wang, S. Zhou, Z. Chen, B. P. Wilson and M. Lundström, *Separation and Purification Technology*, 2021, **259**, 118181.

71 Y. Zhao, B. Liu, L. Zhang and S. Guo, *Journal of Hazardous Materials*, 2020, **396**, 122740.

72 Y. Yu, D. Wang, H. Chen, X. Zhang, L. Xu and L. Yang, *Chemical Research in Chinese Universities*, 2019, **36**, 908–914.

73 M. He, P. Zhang, X. Duan, L. Teng, H. Li, F. Meng, Q. Liu and W. Liu, *Process Safety and Environmental Protection*, 2023, **177**, 1035–1044.

74 J. Lin, L. Li, E. Fan, C. Liu, X. Zhang, H. Cao, Z. Sun and R. Chen, *ACS Applied Materials Interfaces*, 2020, **12**, 18482–18489.

75 J. Lin, C. Liu, H. Cao, R. Chen, Y. Yang, L. Li and Z. Sun, *Green Chemistry*, 2019, **21**, 5904–5913.

76 Y. Chen, P. Shi, D. Chang, Y. Jie, S. Yang, G. Wu, H. Chen, J. Zhu, F. Hu, B. P. Wilson and M. Lundstrom, *Separation and Purification Technology*, 2021, **258**, 118078.

77 Y. Tang, X. Qu, B. Zhang, Y. Zhao, H. Xie, J. Zhao, Z. Ning, P. Xing and H. Yin, *Journal of Cleaner Production*, 2021, **279**, 123633.

78 J. Shi, C. Peng, M. Chen, Y. Li, H. Eric, L. Klemettinen, M. Lundström, P. Taskinen and A. Jokilaakso, *JOM*, 2019, **71**, 4473–4482.

79 J. Xiao, R. Gao, B. Niu and Z. Xu, *Journal of Hazardous Materials*, 2021, **407**, 124704.

80 J. Jegan Roy, M. Srinivasan and B. Cao, *ACS Sustainable Chemistry & Engineering*, 2021, **9**, 3060–3069.

81 İ. B. Pehlivan, R. Marsal, G. A. Niklasson, C. G. Granqvist and P. Georén, *Solar Energy Materials and Solar Cells*, 2010, **94**, 2399–2404.

82 H. Kim, W. Choi, J. Yoon, J. H. Um, W. Lee, J. Kim, J. Cabana and W.-S. Yoon, *Chemical Reviews*, 2020, **120**, 6934–6976.

83 C. Shao, H. Liu, Z. Yu, Z. Zheng, N. Sun and C. Diao, *Solid State Ionics*, 2016, **287**, 13–16.

84 L. Schwich, M. Küpers, M. Finsterbusch, A. Schreiber, D. Fattakhova-Rohlfing, O. Guillon and B. Friedrich, *Metals*, 2020, **10**, 1523.

85 M. Ali Nowroozi, A. Iqbal Waidha, M. Jacob, P. A. van Aken, F. Predel, W. Ensinger and O. Clemens, *ChemistryOpen*, 2022, **11**, e202100274.

5
Direct Regeneration Technologies

Binglei Jiao, Zhichao Yang, Tiyu Jiao, Mengqi Kong, Panpan Xu

5.1 Brief Introduction

With the national "dual carbon" goals gaining prominence, the recycling process of spent cathode materials must evolve toward shorter, more efficient processes with higher added value. As a result, researchers have started focusing on the development of direct regeneration technologies, which are closed-loop recycling method that uses nondestructive physicochemical strategies to heal the structure and compensate for the elements in degraded cathode materials. The electrochemical performance degradation of various commercial materials is mainly due to the deactivation of the active materials in the cathodes. Therefore, spent cathode materials have the potential for direct repair after supplementing lithium and other lost elements. The core advantage of this technology, compared to traditional elemental extraction-based recycling, lies in the protection of crystal structure integrity. This performance-oriented regeneration strategy provides a green and efficient solution for low-carbon recycling of power batteries. Currently, cathode material repair technologies are divided into three categories: solid-state sintering repair technology, liquid-phase hydrothermal repair technology, and electrochemical repair technology.

5.1.1 Direct Regeneration of Spent Cathode Materials

5.1.1.1 Solid-state Regeneration Technology

The solid-state sintering regeneration technology utilizes solid-state reactions between spent cathode materials and lithium sources at high temperatures to replenish elemental deficiencies while concurrently promoting the restoration of their crystal structure. This solid-state sintering regeneration technology is characterized by its short process flow and low environmental impact. Furthermore, it is applicable to various commercial cathode materials currently in use, including lithium cobalt oxide ($LiCoO_2$, LCO), nickel-rich ternary cathodes ($LiNi_xCo_yMn(Al)_{1-x-y}O_2$, NCM/NCA), and lithium iron phosphate ($LiFePO_4$, LFP). This positions the technology as highly competitive for future large-scale industrial recycling applications.

5.1.1.1.1 Solid-state Regeneration of LFP

For olivine-type LFP, the material undergoes phase transformation to ferric phosphate ($FePO_4$) upon extended electrochemical cycling. During battery operation, the formation of a solid electrolyte interphase (SEI) on the anode and a cathode-electrolyte interphase (CEI) on the cathode consumes a portion of the lithium salts during the initial charging process. The development of these interphases leads to irreversible loss of active lithium, as some lithium ions extracted from the LFP lattice cannot be reincorporated. The SEI layer on the anode increases the internal impedance of the cell, which can promote lithium dendrite growth, further depleting active lithium ions and resulting in lithium deficiency within the LFP structure. If the deficient elements remain in the form of salts within the material, direct sintering under an inert atmosphere can effectively restore the crystal structure of LFP. However, for retired materials suffering from elemental loss, it is necessary to introduce additional lithium, iron, and phosphorus sources for elemental compensation.

Various inorganic salts—such as Li_2CO_3, FeC_2O_4, and $NH_4H_2PO_4$—are commonly used as supplement agents to adjust the elemental composition of the waste material back to the stoichiometric ratio. A team led by Professor Guangmin Zhou at Tsinghua University[1] has developed a regeneration method utilizing organic lithium salts, which exhibit unique structural regulation advantages in the repair of degraded cathodes. The abundant active functional groups on their surface, such as cyano and lithium-oxygen groups, form strong chemical coupling with the LFP particle surfaces, enabling effective reconstruction of the material interface. Thermal decomposition studies reveal that organic lithium salts initially decompose at low temperatures to form an amorphous carbon coating on the particles. When the temperature rises above 400 °C, they further release active lithium in the form of Li_2O. This "graded decomposition" behavior ingeniously achieves a synergistic effect of "low-temperature carbon coating for enhanced electron transport and high-temperature lithium compensation for lattice defect repair." Notably, the cyano functional groups reduce iron ions to electrochemically more active ferrous ions through electron transfer during the regeneration process, significantly reducing material polarization. The regenerated LFP (RLFP) demonstrates electrochemical performance that can surpass that of currently commercial LFP cathodes. By tailoring the functional groups and decomposition pathways of the organic lithium salts, this method can be adapted to meet the regeneration requirements of various degraded cathode materials. Compared to conventional high-temperature pyrometallurgy or strong-acid hydrometallurgical processes, this technique reduces the processing temperature by over 40% and produces no harmful by-products, offering significant advantages in energy efficiency, cost-effectiveness, and environmental friendliness. It provides an innovative solution for the solid-phase recycling system of LFP (Figure 5.1).

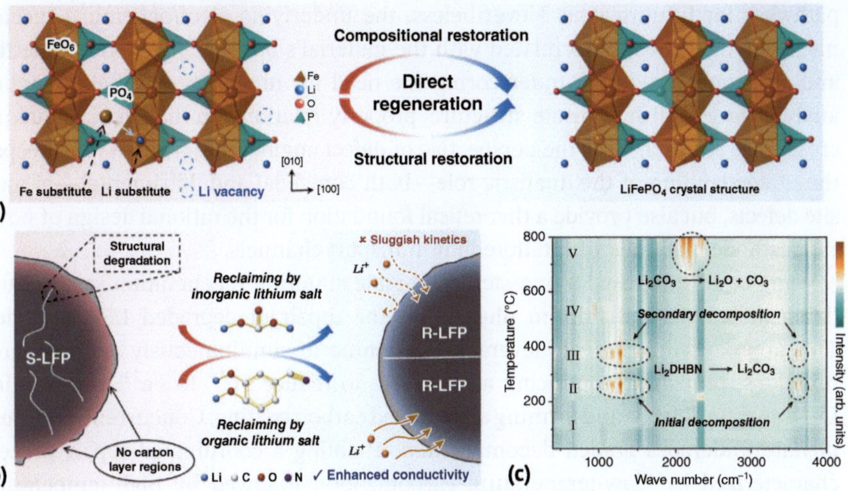

Figure 5.1 (a) Schematic of the crystal structure degradation, highlighting Li-Fe antisite defects and Li vacancies; (b) comparison of regeneration mechanisms using inorganic versus multifunctional organic lithium salts (Li_2DHBN); (c) TG-IR analysis revealing the multistage decomposition process of Li_2DHBN during thermal treatment.
Source: [1] / Springer Nature/ CC BY 4.0.

In the solid-state regeneration process, the presence of Li-Fe antisite defects significantly impacts the electrochemical performance of LFP. Owing to the one-dimensional nature of lithium-ion diffusion pathways in LFP, the incorporation of larger iron atoms into lithium sites can effectively block these diffusion channels.[2, 3] When lithium compensation is performed on materials with a high concentration of Li-Fe antisite defects, lithium-ion intercalation tends to occur preferentially in localized regions, thereby hindering deep and homogeneous lithiation throughout the bulk material.[4, 5] Consequently, it is essential to employ reducing atmospheres—such as hydrogen or carbon monoxide—to mitigate Li-Fe antisite defects. This remediation serves to reopen obstructed lithium diffusion pathways, thereby enabling uniform and deep lithium replenishment.[6] Precise control over the concentration of Li-Fe antisite defects is critical throughout the regeneration process. As an intrinsic defect within the LFP crystal structure, Li-Fe antisite configurations represent a persistent structural feature that arises during both material synthesis and electrochemical cycling. Conventional understanding generally attributes deleterious effects to such cation disorder, positing that it impedes lithium-ion transport kinetics. However, emerging studies reveal that these antisite defects can facilitate the construction of a three-dimensional lithium diffusion network, which enhances anisotropic diffusion characteristics by approximately two orders of magnitude. This defect-enabled mechanism allows for the simultaneous optimization of both rate capability and energy density.[5, 7] Notably, a dynamic coupling mechanism has been identified between antisite defects and lattice tensile strain. Through synergistic stress-defect interactions, this coupling can reconstruct multidirectional migration pathways for lithium ions. Nevertheless, the underlying electrochemical mechanisms are likely deeply correlated with the material's intrinsic magnetic properties and electronic structure, underscoring the need for multi-physical field coupling analyses to establish accurate structure–property relationship models. Conducting cross-scale research from the perspective of defect engineering will not only deepen the understanding of the dualistic role—both beneficial and detrimental—of antisite defects, but also provide a theoretical foundation for the rational design of novel LFP cathode materials with tailored ion-transport channels.

Wang et al.[8] proposed a one-step solid-state sintering regeneration strategy utilizing melamine and lithium phytate for the repair of degraded LFP cathodes. This approach innovatively leverages melamine to simultaneously achieve three functions: providing a reducing atmosphere to reduce Fe^{3+} to Fe^{2+}, remediating Li-Fe antisite defects, and forming an N-doped carbon coating. Concurrently, lithium phytate undergoes graded decomposition, enabling a coordinated repair process characterized by "low-temperature carbonization followed by high-temperature lithium replenishment." Characterization results indicate that the concentration of Li-Fe antisite defects in the regenerated material (RLFP) was significantly reduced from 3 to 0.7%, accompanied by the formation of a uniform N-doped carbon layer on the surface. The regenerated material exhibits outstanding electrochemical performance, delivering an initial discharge capacity of 150.3 mAh/g at 0.2C, maintaining 82% capacity retention after 300 cycles at 1C, and demonstrating high-rate capability at 5C comparable to that of commercial LFP (Figure 5.2).

During extended charge-discharge cycling, the carbon coating on LFP surfaces may deteriorate due to oxidation, mechanical stress, or parasitic side reactions,

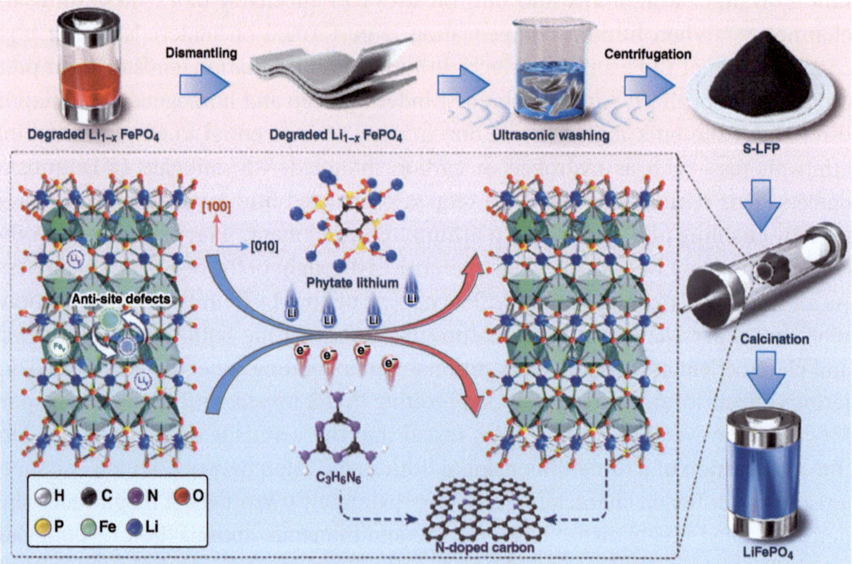

Figure 5.2 Schematic illustration of the degradation mechanism and regeneration process of spent LFP. *Source:* [8] / with permission of John Wiley & Sons.

resulting in diminished electrical conductivity and elevated interfacial impedance, which consequently impair overall battery performance. Therefore, carbon removal represents an essential procedure in the material regeneration workflow to reclaim its electrochemical activity.[9] An effective decarburization strategy must eliminate the dysfunctional carbon layer without compromising the structural integrity of the LFP olivine crystal lattice, while also facilitating the subsequent application of a uniform secondary carbon coating. The principal technical routes for decarburization can be categorized into three groups: physical exfoliation,[10] chemical oxidation, and high-temperature pyrolysis. Physical exfoliation techniques directly detach the surface carbon layer through mechanical action, with ball milling and jet milling serving as representative methods. Ball milling operates by applying collision and shear forces via grinding media to exfoliate the carbon coating; successful implementation, however, requires precise regulation of rotational speed and processing duration to prevent excessive comminution, which can induce particle fracture or lattice damage. Jet milling achieves carbon removal through high-velocity gas stream impacts, offering superior efficiency at the cost of more demanding equipment specifications. While physical methods are operationally straightforward and economically favorable, they often prove inadequate for complete elimination of carbon residues trapped within material pores and may incidentally introduce secondary contamination. Chemical oxidation approaches leverage strong oxidizing agents—such as concentrated sulfuric acid, nitric acid, hydrogen peroxide, or ozone—to selectively etch carbon. For example, concentrated sulfuric acid carbonizes organic components under heated conditions, and subsequent oxidation with hydrogen peroxide converts carbon into CO_2 and H_2O, thereby dissolving the carbon layer. This method delivers high carbon removal efficiency but necessitates stringent control over reaction parameters—including temperature, concentration, and duration—to avert acid corrosion of the LFP bulk material or oxidation of Fe^{2+} to Fe^{3+}, either of which would accelerate capacity fade. Furthermore, chemical processes typically generate liquid waste streams, requiring dedicated effluent treatment systems. High-temperature pyrolysis entails heating the material to 500–800 °C within an inert atmosphere, thermally decomposing the carbon coating into CO/CO_2 gases that are subsequently evolved. Temperature selection is critical and must balance carbon removal efficiency against material thermal stability: processing below 600 °C frequently yields substantial carbon residue, whereas temperatures exceeding 750 °C risk decomposing LFP into secondary impurity phases such as Fe_2P. Process efficiency can be enhanced by introducing gasifying agents like steam or CO_2 to promote carbon gasification reactions. Although pyrolysis is a mature and chemically clean technology, it remains energy-intensive and requires careful process control to suppress particle agglomeration caused by high-temperature sintering. Wang et al.[11] developed a regeneration strategy combining oxidative roasting with solid-state reduction for the reclamation of degraded LFP material. By optimizing the oxidation temperature to 500 °C, the study achieved a 90.5% carbon removal rate, effectively addressing the performance degradation caused by carbon accumulation in conventional regeneration processes. Under optimal conditions (500 °C oxidation followed by 700 °C reduction), the regenerated material demonstrated excellent electrochemical performance, achieving a discharge capacity of 152.3 mAh/g at 0.1C and maintaining 130.7 mAh/g after 100 cycles at 1C (Figures 5.3 and 5.4).

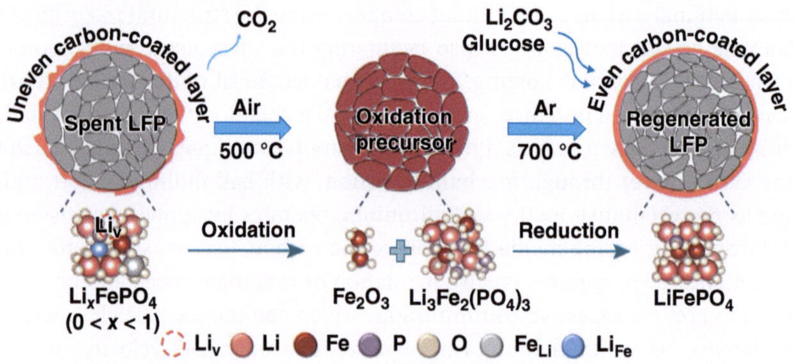

Figure 5.3 Schematic illustration of the oxidation roasting and solid-phase reduction process for regenerating spent LFP materials. *Source:* [11] / with permission of Elsevier.

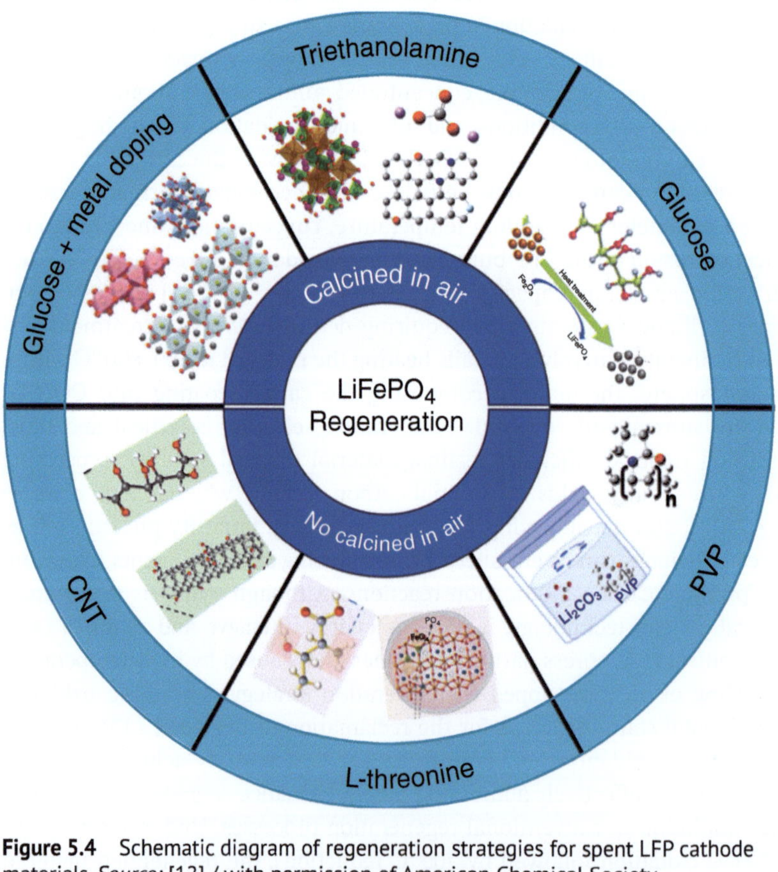

Figure 5.4 Schematic diagram of regeneration strategies for spent LFP cathode materials. *Source:* [12] / with permission of American Chemical Society.

Current direct regeneration technologies for LFP have evolved along two primary technical pathways. The non-atmospheric firing route employs methods such as carbon nanotube modification, hydrothermal repair with amino acids, and polyvinylpyrrolidone (PVP) coating to enhance performance, yet faces limitations in achieving complete regeneration. In contrast, the atmospheric firing route enables thorough regeneration through preliminary impurity removal. Specific implementations include structural reconstruction via glucose-assisted repair, optimization of interfacial transport using glucose/titanium dioxide composite systems, and synergistic defect remediation with nitrogen doping through melamine-lithium phytate systems. Building upon the atmospheric firing approach, Ma et al. innovatively utilized triethanolamine as a multifunctional repair agent. Following impurity removal at 600 °C in air, triethanolamine simultaneously enabled Fe^{3+} reduction, elimination of Li-Fe antisite defects, and construction of a nitrogen-doped carbon coating. This method completely avoids the use of acids or alkalis while achieving a specific capacity of 152.87 mAh/g at 0.2C (95% of commercial LFP) and maintaining 88.52% capacity retention after 600 cycles at 1C. The strategy realizes synergistic optimization of structural repair and interface modification, providing an efficient regeneration solution for spent LFP that combines outstanding electrochemical performance with environmental benignity.[12]

Challenges of solid-state regeneration of LFP: Solid-state regeneration technology primarily relies on high-temperature calcination to restore material properties, yet it faces core issues related to difficult process control and inconsistent performance of the regenerated products. The solid-state method involves mixing spent LFP with lithium sources, carbon additives, and other supplements, followed by high-temperature treatment. However, the calcination process is prone to complications such as lithium volatilization, particle agglomeration, and the formation of impurity phases. For instance, the calcination temperature requires precise control—insufficient temperatures lead to incomplete lithium replenishment, whereas excessive temperatures may induce reduction of iron species, generating impurities and compromising the material structure. Furthermore, spent LFP materials originate from diverse sources with variations in morphology, particle size, and degradation mechanisms, making it challenging for conventional solid-state sintering to achieve uniform regeneration across different batches of waste materials.

Future directions of solid-state regeneration of LFP: Solid-state regeneration represents a crucial methodology for reclaiming spent LFP batteries, primarily achieving structural restoration and performance recovery through high-temperature calcination. The calcination temperature, however, remains a critical determinant of final material performance. Inadequate temperatures result in insufficient lithium replenishment, failing to effectively restore electrochemical properties, while excessively high temperatures can trigger reduction of iron species, generating impurity phases that compromise structural integrity. Consequently, the development of intelligent temperature control systems coupled with online monitoring technologies emerges as pivotal for achieving precise thermal management. Real-time monitoring and dynamic regulation of calcination parameters can effectively suppress lithium volatilization and iron reduction, thereby enhancing performance consistency in regenerated materials. Beyond temperature control, the selection and proportion of additives significantly influence regeneration outcomes. Lithium sources and carbon additives participating in the high-temperature process not only facilitate lithium compensation but also critically impact grain boundary structure and electrochemical characteristics. Future efforts should focus on developing multifunctional composite additives capable of simultaneously delivering lithium replenishment, grain boundary blocking, and impurity adsorption under thermal conditions. The introduction of highly stable coating agents can effectively mitigate particle agglomeration and impurity phase formation, ultimately enhancing the electrochemical performance of regenerated products. Furthermore, the heterogeneous nature of spent LFP materials—exhibiting variations in morphology, particle size distribution, and degradation mechanisms across different batches—poses significant challenges for achieving uniform regeneration through conventional solid-state sintering. Establishing comprehensive material databases paired with process-matching models, supported by intelligent algorithms to predict optimized sintering parameters for diverse waste streams, will enable precise process adaptation. Simultaneously, developing universal regeneration protocols applicable to varied feedstock represents another critical research direction that would substantially enhance both the applicability and economic viability of solid-state regeneration technology. Regarding equipment innovation, traditional batch-processing systems demonstrate limited suitability for large-scale industrial implementation. Future advancements should prioritize the development of continuous calcination equipment through optimized heating mechanisms and material transport systems, ultimately improving production efficiency and energy utilization in commercial-scale regeneration processes.

5.1.1.1.2 Solid-state Regeneration of LCO

In the domain of solid-state regeneration, the solid-phase repair method for LCO represents an efficient and practical strategy for cathode reclamation. This approach, which shares similarities with the conventional cathode synthesis process, effectively restores both the original composition and the crystal structure of degraded materials. Typically, the spent cathode powder (e.g., $Li_{1-x}CoO_2$) is mixed with a lithium supplement salt—such as Li_2CO_3, $LiOH{\cdot}H_2O$, or $LiNO_3$—in a stoichiometric ratio. The mixture is subsequently subjected to high-temperature sintering under an air or oxygen atmosphere, achieving the dual objectives of lithium replenishment and lattice defect remediation. Precise control over the Li/transition metal (Li/TM) molar ratio is essential to obtain regenerated products with satisfactory electrochemical performance. Recently, a team from Tsinghua University led by Wang et al.[13] proposed a sustainable regeneration strategy to upcycle spent LCO (SLCO) into high-voltage cathode materials. They combined the SLCO with Al_2O_3, MgO, and Li_2CO_3 in a specific stoichiometric ratio and successfully produced a Mg/Al dual-doped, upgraded LCO cathode via a single-step solid-state reaction. The regenerated material exhibits a smooth surface and a well-ordered layered structure, delivering exceptional electrochemical performance at a high charging voltage of 4.6 V, retaining 79.7% of its initial capacity after 300 cycles. Notably, this direct regeneration method offers substantial economic and environmental advantages compared to traditional metallurgical recovery processes, which are typically energy-intensive and generate significant pollution. A techno-economic analysis indicates that regenerating one ton of SLCO batteries using this approach can yield a profit of up to $3,671.3. Ji et al.[14] developed a direct regeneration strategy for SLCO that employs LiF as both a "crack/defect healing agent" and a "structural stabilizer." During high-temperature sintering, LiF not only effectively fills and heals the intraparticle cracks and surface defects induced by long-term cycling but also reconstructs the material interface through gradient doping of F atoms, forming a F-rich surface layer. This layer significantly suppresses parasitic reactions between highly oxidized Co^{4+}/lattice oxygen (O^-) species and the electrolyte, thereby enhancing the stability of lattice oxygen. Theoretical calculations further demonstrate that the incorporation of F increases the energy level difference between Co 3d and O 2p orbitals, effectively inhibiting oxygen loss and cobalt dissolution. The regenerated U-LCO material exhibits exceptional cycling stability and a high lithium-ion diffusion coefficient at a cutoff voltage of 4.5 V, achieving a capacity retention of 81.6% after 260 cycles—surpassing that of commercial pristine LCO.

Liu et al.[15] developed a "killing three birds with one stone" solid-state sintering strategy that enables direct upcycling of degraded $LiCoO_2$ (D-LCO). The innovation of this approach lies in its simultaneous achievement of three critical functions through a simple one-step sintering process: lithium replenishment, Li_2SO_4 surface coating, and multielement doping with Mn, N, and S. Their investigation reveals that Mn preferentially occupies Co sites while N and S incorporate into the interlayer spaces of Li-O slabs. This synergistic effect, combined with the protective Li_2SO_4 coating, collectively stabilizes the material's crystal structure under high-voltage operation, effectively suppressing lattice oxygen escape and detrimental phase transitions. The regenerated and upgraded material (MNS-LCO) demonstrates exceptional electrochemical performance at a high cutoff voltage of 4.5 V, delivering a remarkable initial discharge capacity of 188.2 mAh/g and capacity retention rates of 92.5% after 100 cycles and 86.4% after 300 cycles at 0.5C—surpassing the performance of pristine commercial LCO materials. This streamlined process establishes a new paradigm for transforming degraded LCO (D-LCO) into high-energy-density, long-lifespan cathode materials.

Zeng et al.[16] innovatively proposed a strategy for direct upcycling of SLCO/graphite/copper mixtures. The core of this research lies in developing a "bulk-oriented doping" process that ingeniously utilizes both the thermal energy released from graphite oxidation in the waste materials and the significantly reduced migration energy barrier for copper (from 0.57 to 0.55 eV), facilitated by abundant lithium vacancies in the material. This enables successful incorporation of copper impurity atoms into the LCO bulk phase at a relatively low temperature of 700 °C. Through optimization of copper content (0.7 wt.%), dual-site anchoring of copper at both Li and Co sites was achieved, forming a unique reverse concentration gradient distribution characterized by lower surface and higher internal concentrations. This distinctive doping behavior effectively stabilizes the crystal structure, expands interlayer spacing, and lowers the O 2p-band center, thereby suppressing lattice oxygen release. The optimized regenerated material (RLCO-0.7Cu) demonstrates exceptional electrochemical performance at a high voltage of 4.5 V, delivering an initial discharge capacity of 177.59 mAh/g at 0.2C, with capacity retention rates of 89.1% after 200 cycles at 1.0C, and maintaining 154.8 mAh/g with 82.4% retention after 500 cycles even at a high rate of 5.0C (Figure 5.5).

Challenges of solid-state regeneration of LCO: The development of LCO regeneration technologies holds significant importance for environmental protection, resource conservation, and sustainable development. However, current regeneration approaches—spanning solid-state, liquid phase, and emerging technical routes—exhibit substantial limitations. In solid-state regeneration, although mechanical mixing and high-temperature calcination can reconstruct the material structure, the process fundamentally requires substantial energy input to maintain elevated temperatures (typically exceeding 800 °C). Furthermore, lithium volatilization under these conditions leads to deviations from stoichiometric composition (typically requiring 5–10 wt.% excess lithium compensation), resulting in low regeneration efficiency and elevated costs that hinder their applicability to large-scale recycling operations.[17–19] Additionally, high-temperature processing can induce irreversible damage to the layered structure of $LiCoO_2$,

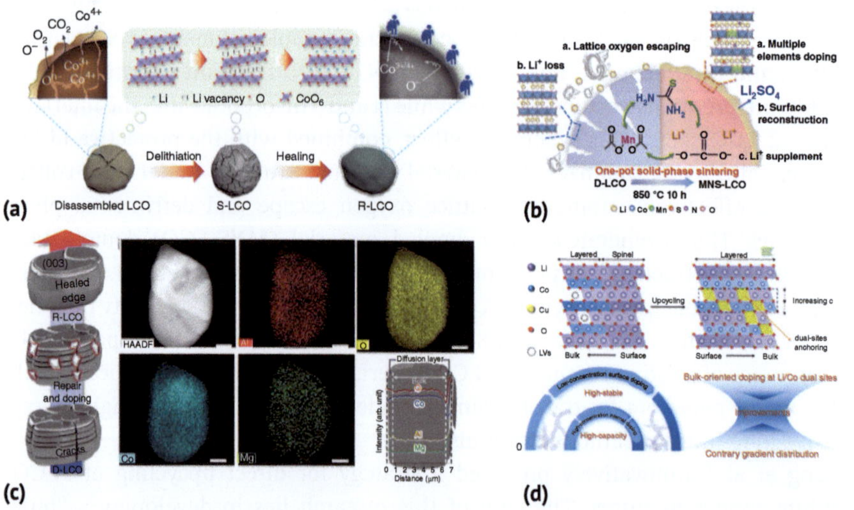

Figure 5.5 (a) Schematic illustration of the healing process for SLCO using LiF as a repair agent. *Source:* [14] / with permission of Elsevier. (b) Schematic of the "one-pot solid-phase sintering" upcycling process for D-LCO. *Source:* [15] / with permission of John Wiley & Sons. (c) High-angle annular dark-field scanning transmission electron microscopy (HAADF-STEM) image and elemental distribution of R-LCO. *Source:* [13] / with permission of Springer Nature. (d) Schematic of bulk-oriented Cu doping and contrary gradient distribution in upcycled LCO. *Source:* [16] / with permission of Elsevier.

triggering lattice distortion and Li/Co cation disordering.[20] The inherently slow kinetics of solid-state reactions also cause inhomogeneous distribution of replenished elements, leading to localized lithium vacancy accumulation and incomplete phase transformation from spinel to layered structures in regenerated materials. These defects severely compromise ionic diffusion rates and electronic conductivity, yielding materials with poor morphological uniformity and unstable electrochemical performance.[21, 22]

Future directions of solid-state regeneration of LCO: Solid-state regeneration serves as a fundamental methodology in the recycling of lithium-ion battery (LIB) cathode materials. Beyond achieving synergistic recovery through lithium replenishment and structural repair, its advantages include operational simplicity, minimal environmental impact, and significant potential for scalable production, contributing to its widespread application in battery recycling. However, conventional solid-state approaches typically rely on high-temperature solid-state reactions to reconstruct the $LiCoO_2$ structure. The reaction kinetics are inherently limited by inefficient solid-solid interfacial contact, resulting in slow and inhomogeneous lithium-ion diffusion alongside prolonged processing durations. Furthermore, inadequate mixing frequently induces localized Li/Co ratio imbalances, substantially compromising the integrity of the layered structure. The high-temperature calcination may also trigger excessive grain growth, reducing specific surface area, and consequently impairing rate capability. To address these challenges, auxiliary techniques such as high-energy ball milling and pulverization may be implemented to reduce particle sizes of both the precursor and lithium source, thereby enhancing interfacial contact area and mixing homogeneity. From an economic perspective, the high calcination temperatures and extended reaction times characteristic of solid-state methods lead to substantially increased energy consumption. Therefore, reducing energy requirements and operational costs remains crucial for further development. Promising directions include developing low-temperature, short-process solid-state technologies—such as introducing mechanochemical activation to lower calcination temperatures—or implementing rapid heating methods like Joule heating to significantly shorten processing time.

5.1.1.1.3 Solid-state Regeneration of NCM Cathodes

In the late 1990s, NCM ternary composite materials were first proposed, leading to the synthesis of various $LiNi_xCo_yMn_zO_2$ (NCM) compositions with layered lithium TM oxide structures analogous to LCO. Common formulations include NCM111, NCM523, NCM622, and NCM811. With the rapid expansion of new energy vehicles and energy storage markets, NCM LIBs have been widely adopted due to their high energy density, excellent cycling performance, and substantial specific capacity. However, after prolonged use, NCM cathode materials frequently undergo performance degradation through structural deterioration, cation disordering, and diminished electrochemical activity, generating substantial quantities of spent batteries.[23, 24] Compared to LCO and LFP batteries, the multimetallic composition of

NCM materials significantly complicates the recycling process.[25] Consequently, the development of efficient regeneration technologies has emerged as a critical research focus. Rational regeneration strategies can effectively restore the crystalline structure and electrochemical functionality of degraded materials, enabling resource circularity while reducing raw material consumption and environmental impact (Figure 5.6).

Jin et al.[26] proposed a defect-driven prelithiation strategy for direct regeneration of degraded NCM cathodes. Their innovative approach utilizes mechanical impact to induce surface oxygen defects, which significantly enhances LiOH adsorption on NCM surfaces and reduces the kinetic energy barrier for Li^+ migration. By modulating the local thermodynamic states (temperature and pressure), topotactic prelithiation was achieved, thereby facilitating the rapid and ordered restoration of the layered structure during subsequent high-temperature sintering. The regenerated NCM material demonstrates exceptional electrochemical performance, maintaining 90% capacity retention after 200 cycles and even surpassing the rate capability of commercial pristine NCM. This study provides new insights into green and efficient regeneration of NCM cathodes, revealing the crucial role of synergistic modulation between defect kinetics and thermodynamics in cathode regeneration.

In recent years, driven by growing demands for battery recycling, domestic research teams have achieved significant breakthroughs in the development and process optimization of regeneration technologies for NCM cathode materials. Particularly, the exploration of solid-state regeneration has established a robust foundation for enhancing both the efficiency and performance recovery of NCM materials. Solid-state regeneration typically involves mixing the cathode powder with a lithium salt—to replenish lost lithium ions—in a specific ratio, followed by calcination under an air or oxygen atmosphere at a predetermined temperature for a set duration to obtain the final regenerated product.[27, 28] In terms of

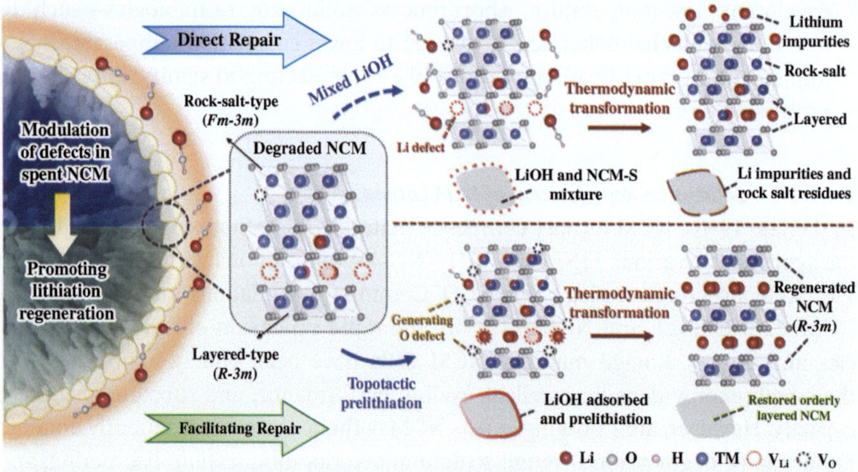

Figure 5.6 Schematic diagram of defect regulation and direct regeneration mechanism for spent NCM cathodes. *Source:* [26] / with permission of John Wiley & Sons.

specific technological advances, a team led by Professor Ge Peng from Central South University[27] proposed a homogeneous regeneration strategy for NCM523, combining physical crushing with solid-state sintering. This method not only effectively repaired structural defects such as oxygen vacancies and Li/Ni cation mixing but also yielded regenerated material with uniform single-crystal particle size and smooth surfaces, significantly enhancing both electrochemical performance and structural stability. Separately, Fan et al.[28] successfully regenerated spent NCM523 cathode material by calcination at 800 °C in air. The regenerated cathode delivered a discharge specific capacity of 141.6 mAh/g after 100 cycles, performing comparably to commercial benchmarks. The repaired material exhibited favorable Li^+ transport kinetics and minimal polarization (Figure 5.7).

Notably, certain recycled cathode materials can exhibit superior electrochemical performance and enhanced value compared to their pristine counterparts through direct recycling processes, thereby achieving "upcycling." Among these approaches, the conversion of spent polycrystalline NCM into single-crystal NCM has emerged as a prominent research focus, with the eutectic salt method playing a crucial role as a versatile and efficient technique in this upgrading process. For instance, a team led by Professor Chen Gen from Central South University[31] proposed a direct regeneration strategy based on a stable eutectic molten salt system. By optimizing the KCl-LiCl-LiOH molten salt composition, they successfully transformed degraded polycrystalline NCM811 cathode material into homogeneous single-crystal structures while effectively replenishing the lost active lithium. This method significantly enhanced the structural stability and high-voltage tolerance of the regenerated cathode material. Similarly, the team successfully achieved direct regeneration of degraded polycrystalline NCM523 cathodes by constructing a $LiOH$-$LiNO_3$-CH_3COOLi ternary molten salt system.[32] The core of this strategy lies in utilizing the low-temperature fluidity and high ionic conductivity of the molten salt to achieve uniform, bulk-phase lithium replenishment at 400 °C. Subsequent thermal treatment effectively eliminated cation disordering, restoring the material to a level comparable to commercial pristine NCM523, as evidenced by a capacity

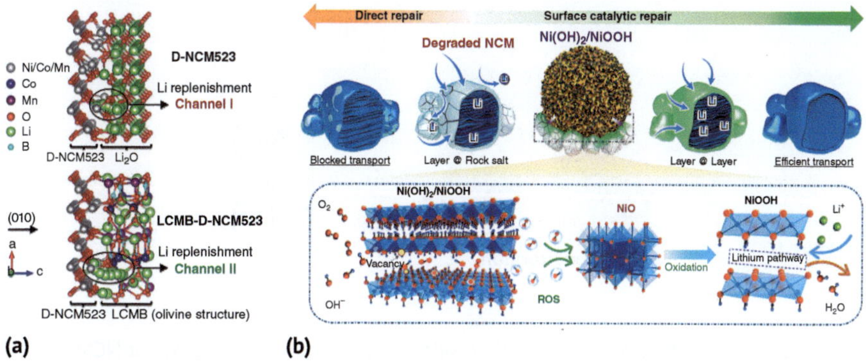

Figure 5.7 (a) Schematic of fast Li replenishment channel construction. *Source:* [29] / with permission of John Wiley & Sons. (b) Schematic of surface catalytic repair mechanism. *Source:* Adapted from [30].

retention of 93.7% after 100 cycles at 0.5 °C. Furthermore, addressing the challenge of high lithium diffusion barriers in degraded NCM cathodes—where surface rock-salt/spinel phases lack effective pathways for lithium replenishment—Professor Zhou Guangmin's team at Tsinghua University[29] developed an assisted repair technology utilizing rapid lithium transport channels. By constructing $LiCo_{0.5}Mn_{0.5}BO_3$ channels, they substantially reduced the energy barrier for lithium incorporation, shortened the regeneration time to merely 2 hours, and improved both the thermal stability and electrochemical performance of the material. This technology not only reduced lithium source consumption to just 25% of conventional methods but also significantly raised the material's thermal decomposition temperature, enabling stable cycling over 200 cycles even at 60 °C. Concurrently, the team has been actively exploring novel technical pathways such as surface catalytic repair.[30] This technique leverages the surface catalytic activity of TM hydroxides to convert rock–salt phases back into layered structures, dramatically improving lithium-ion migration rates and regeneration efficiency. By introducing polycrystalline $Ni_{0.5}Co_{0.2}Mn_{0.3}(OH)_2$ as a surface catalyst, they successfully regenerated degraded layered oxide cathode materials, with the repaired material maintaining 88.4% capacity retention after 1,000 cycles in 1.3 Ah pouch cells. These technological advances not only improve recycling efficiency but also open new pathways for the regeneration of spent battery materials, providing crucial support for the effective recycling and reuse of retired NCM batteries (Figure 5.8).

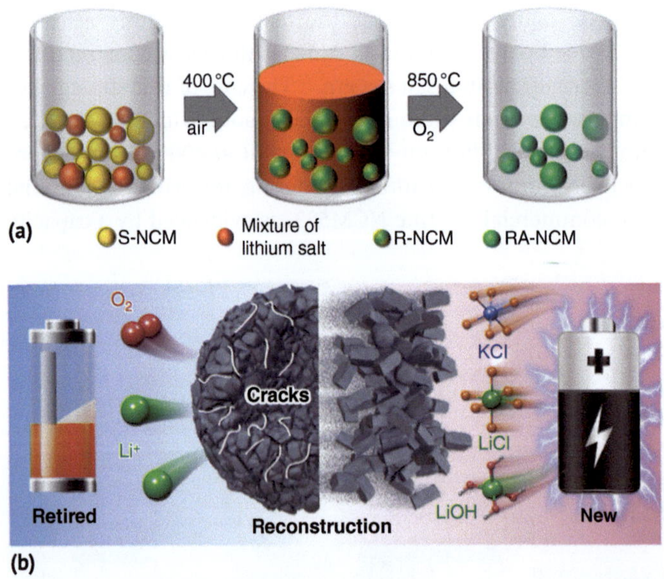

Figure 5.8 Schematic diagrams of molten salt-assisted regeneration for spent NCM cathodes. (a) Process of ternary molten salt (e.g., $LiOH\text{-}LiNO_3\text{-}CH_3COOLi$) relithiation and annealing. *Source:* Adapted from [32]. (b) Microstructural reconstruction from polycrystal to single crystal with Li replenishment. *Source:* [31] / with permission of John Wiley & Sons.

Challenges of solid-state regeneration of NCM: The regeneration of NCM cathode materials currently faces multiple challenges across different technical routes. While solid-state regeneration offers economic efficiency and can restore degraded materials through relatively simple thermal processing, it requires precise determination of the Li/TM molar ratio based on the atomic stoichiometry of the original structure, followed by critical thermal treatment steps to achieve necessary elemental replenishment. Furthermore, the solid-solid interface between the degraded NCM material and the introduced chemical reagents is inherently inhomogeneous, compromising the regeneration effectiveness and hindering the achievement of highly efficient and consistent recovery.[25, 28, 33] When employing the eutectic molten salt method, the sintering process typically necessitates excessive lithium salts. Subsequent procedures then wash away most of these molten salts from the cathode product, resulting in significant lithium salt waste and creating a demand for additional recovery processes to mitigate losses.[34] Liquid-phase regeneration, though not requiring precise lithium dosage control, involves high-pressure operations that present substantial safety risks. Moreover, a growing body of research indicates that hydrothermally regenerated materials often lack complete structural stability, typically necessitating an additional brief calcination step post-hydrothermal treatment. This requirement significantly diminishes the economic viability of hydrothermal recycling for NCM and other cathode materials, particularly since comparable regeneration can often be achieved via solid-state sintering or solid-solution treatment followed by high-temperature calcination. Emerging electrochemical regeneration techniques utilize a system combining cathode materials with an inert electrode and a lithium-containing aqueous electrolyte. Under an applied electric field, lithium ions directionally migrate into the layered cathode structure, accomplishing relithiation. However, this method also encounters several technical bottlenecks: the regenerated cathode material requires subsequent processing steps, including current collector removal and active material washing, while preexisting crystal structure damage remains incompletely repaired. Due to these technical limitations, electrochemical regeneration has not yet reached maturity for practical application in recycling degraded NCM cathodes, and its industrialization continues to face significant challenges.

Future directions of solid-state regeneration of NCM: Current regeneration technologies face persistent challenges in managing the heterogeneous composition of end-of-life NCM materials and accurately quantifying lithium loss, making it difficult to ensure consistent performance in regenerated products. Thus, by addressing existing technical limitations and analyzing future requirements, the development of more efficient, economically viable, and environmentally benign regeneration technologies for degraded NCM cathodes constitutes a crucial step toward achieving sustainable development within the lithium battery industry.

Solid-state regeneration stands as one of the primary methods for recycling ternary LIBs. However, ternary cathode materials (NCM) exhibit multi-series characteristics with significant variations in crystalline structure, doping elements, and modification strategies. Current research predominantly focuses on regenerating specific types or a limited category of cathode materials, while different ternary compositions require distinct temperature profiles, durations, and atmospheric conditions during high-temperature regeneration. Therefore, achieving homogeneous crystalline phase and stable performance in regenerated ternary materials across diverse end-of-life compositions represents a critical research objective for the next phase. Furthermore, molten-salt-assisted regeneration faces additional challenges related to the difficulty and cost associated with salt removal. Consequently, solid-state regeneration technologies should advance toward low-temperature rapid sintering and precise lithium compensation strategies, while developing cost-effective, recyclable molten-salt systems. These advancements would collectively reduce energy consumption and salt loss during regeneration, while enhancing the electrochemical consistency of regenerated materials.

5.1.2.1 Liquid-phase Regeneration Technology

Liquid-phase regeneration can be categorized into two types. The first type is based on the hydrothermal chemical reaction of Li^+ in a liquid medium.[35] In a sealed reactor, the high diffusion rate of the high-temperature and high-pressure fluid drives Li^+ to migrate into the interior of the cathode material, simultaneously repairing material defects.[36] Unlike solid-phase sintering regeneration, the amount of lithium source added in liquid-phase regeneration does not need to strictly follow the stoichiometric ratio, making it applicable to retired cathode materials with varying degrees of lithium deficiency.[37] The second type involves room-temperature lithiation regeneration based on topochemical reactions.[38] For cathode materials with a low degree of degradation and intact lithium diffusion channels, spontaneous lithiation can be achieved through the free diffusion of Li^+ in lithium salt solutions at room temperature, thereby avoiding the additional energy consumption associated with high-temperature hydrothermal processes.

5.1.2.1.1 Liquid-phase Regeneration of LFP

Liquid-phase regeneration of LFP involves the introduction of a lithium source and retired LFP material into a liquid solvent. Under high temperature and pressure, Li^+ is reinserted into the degraded LFP structure. Compared with solid-phase sintering regeneration, the lithium source in the hydrothermal method is supplied in excess, eliminating the need to adjust elemental ratios according to strict stoichiometry. Therefore, the hydrothermal method is widely applicable to LFP cathode materials with varying degrees of lithium loss.[39] During the liquid-phase regeneration process, molecules with reducing capabilities promote the targeted repair of Li-Fe antisite defects and eliminate ferric phosphate phases.[40] Therefore, the selection of an appropriate reducing agent is crucial. Common reducing agents include citric acid,[37] sodium sulfite, and natural amino acids.[41] Moreover, the liquid-phase regeneration method is highly versatile, capable of regenerating LFP with varying degrees of lithium deficiency.[42] Furthermore, during battery cycling, the extraction of lithium ions from the LFP lattice may lead to the migration of iron ions to lithium sites, resulting in the formation of Li-Fe antisite defects. Modifying the chemical environment of iron atoms to enhance the strength of Fe–O bonds can effectively suppress iron ion migration, thereby improving the cycling life of LFP.[43] In the liquid-phase regeneration process, functional reducing agents, such as PVP, can be employed to construct a heterogeneous interface between LFP particles and an N-doped carbon layer.[44] The N in the carbon layer alters the chemical environment of the iron ions, significantly increasing the bond energy of the Fe–O bonds. Moreover, tea polyphenols serve as electron donors to reduce Fe(III) phase and heal antisite defects in the hydrothermal process and introducing a hybrid $AlPO_4/Li_3PO_4$ (AP/LP) patching coating at the fracture of the carbon layer on LFP for targeted surface reconstruction, accompanied by Al doping into the bulk LFP. AP/LP patching coating forms a dual electron- and ion-conducting layer with the residual carbon layer, coupling with strengthened Fe–O bonding by Al doping, to improve kinetics and stability of LFP during cycling (Figure 5.9).[45]

Challenges of liquid-phase regeneration of LFP: Liquid-phase regeneration technology reconstructs the material structure via reducing molecules in a liquid medium, with representative methods including hydrothermal regeneration. However, the technique faces two major bottlenecks: complex reaction conditions and environmental concerns. The hydrothermal method requires

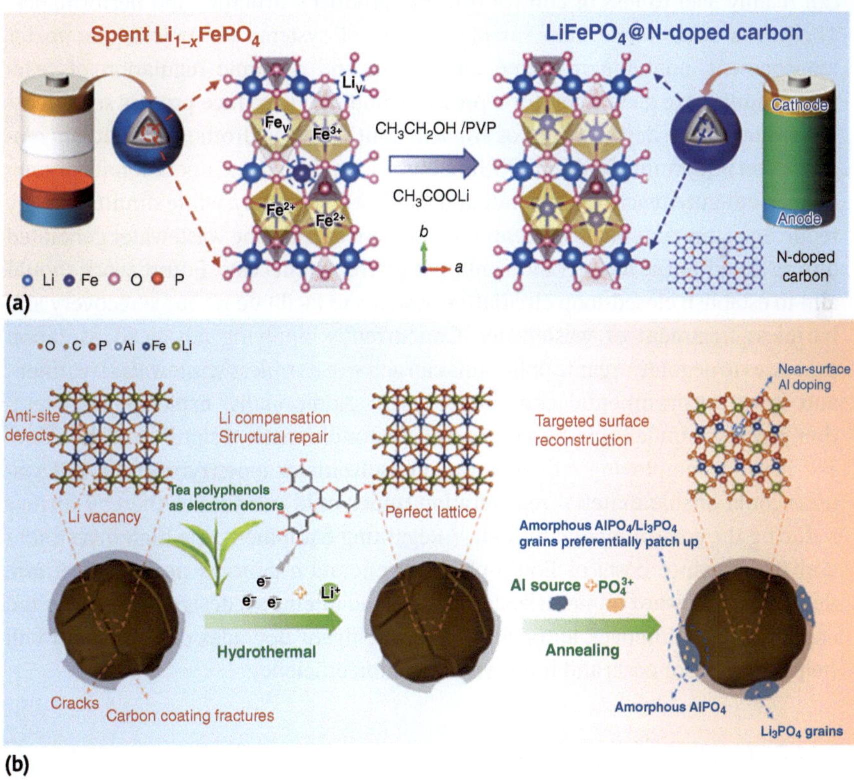

Figure 5.9 Schematic of the liquid-phase regeneration process of LFP using different reducing agents: (a) PVP. *Source:* [44] / with permission of John Wiley & Sons; (b) Tea polyphenols. *Source:* [45] / with permission of John Wiley & Sons.

a high-temperature and high-pressure sealed environment, resulting in high equipment investment and maintenance costs.[42] Moreover, minor fluctuations in parameters such as pH and temperature can easily lead to loss of control over the structure and performance of the final product.[46] Furthermore, the subsequent washing process in liquid-phase regeneration is cumbersome, and wastewater treatment costs increase significantly. More critically, the waste liquids generated from liquid-phase reactions—containing fluorine and acids—require complex neutralization treatments.[47] The associated environmental compliance costs further undermine its economic feasibility.

Future directions of liquid-phase regeneration of LFP: Although the liquid-phase method delivers excellent performance recovery, its process complexity and environmental burden hinder its application in large-scale industrialization. The acids, alkalis, and other chemical reagents used in the liquid-phase regeneration process impose significant environmental pressure, particularly due to the high treatment costs associated with fluorine and acid-containing waste liquids. Future efforts should focus on developing green solvent systems to reduce the generation of hazardous waste. The hydrothermal method requires a high-temperature, high-pressure, sealed environment, and minor fluctuations in reaction conditions can readily lead to loss of control over the product's structure and performance. There is a need to develop intelligent control systems for precise parameter management, enabling real-time monitoring and dynamic regulation of reaction conditions (e.g., temperature, pressure, and pH) to reduce process sensitivity. Furthermore, the development of efficient continuous hydrothermal reactors represents an important direction. Optimizing reactor design and operational parameters can enhance reaction efficiency and product consistency, while simultaneously reducing equipment investment and maintenance costs. The wastewater generated during liquid-phase regeneration entails high treatment costs. Future work should aim to establish closed-loop circulation systems to facilitate resource recovery and harmless treatment of wastewater. Concurrently, applying advanced oxidation processes to degrade organic pollutants can achieve harmless wastewater treatment and lower environmental compliance costs. Additionally, exploring regeneration processes under acid-free or weak-acid conditions is a significant direction. For instance, employing reductive organic solvents or bioenzyme-catalyzed systems could enable material regeneration under mild conditions, thereby further reducing the environmental burden.[45] Regarding equipment, the high investment and maintenance costs of liquid-phase regeneration systems necessitate future development toward miniaturized, modular, and intelligent designs to lower initial capital outlay. Moreover, automation and intelligent upgrades of equipment will help reduce labor costs and improve production efficiency.

5.1.2.1.2 Liquid-phase Regeneration of LCO

Liquid-phase regeneration of LCO is based on the hydrothermal chemical reaction of Li^+ in a liquid medium. Within a sealed reactor, the high diffusion rate of high-temperature and high-pressure fluids drives Li^+ into the cathode material, simultaneously repairing material defects. A lithium-rich solution is generally required to ensure a favorable liquid environment without causing significant alterations to the material's structure. Owing to its uniform deposition, the liquid-phase regeneration process enables the production of homogeneous repaired materials and allows for the tailoring of material microstructures.

Fei et al.[48] proposed an oxidizing hydrothermal reaction to regenerate SLCO. This method operates at lower temperatures and pressures compared to conventional hydrothermal reactions. Hydrogen peroxide and ammonium persulfate were employed as oxidizing agents in the hydrothermal solution. Oxidizing hydrothermal reaction. The hydrothermal oxidation process oxidized Co^{2+} to Co^{3+} in the SLCO, and the regenerated LCO (R-LCO) exhibited a single-crystal morphology. Conventional liquid-phase repair is limited to replenishing Li^+ and oxidizing Co^{2+} via high-pressure hydrothermal reactions, and fails to address the dissolution and loss of cobalt. Wang et al.[49] utilized a $LiCl–CH_4N_2O$ deep eutectic solvent (DES) to regenerate SLCO under ambient pressure. In contrast to aqueous solutions, the liquid urea-based DES exhibits a higher adsorption energy for Co^{2+} (−1.82 eV) than for Li^+ (−0.89 eV). This stronger interaction promotes the preferential diffusion of Co^{2+} from the DES into SLCO. By adjusting the molar ratio of LiCl to CH_4N_2O, the eutectic point of the DES can be lowered to below 120 °C. This reaction temperature is lower than that of most conventional hydrothermal regeneration processes. The R-LCO delivered a capacity of 133.1 mAh/g at 0.1C. Moreover, the DES could be recovered with a high recycling rate of 90%.

The R-LCO through liquid-phase regeneration followed by a short annealing process suffers from surface instability under high-voltage conditions. To address this issue, Wang et al.[50] developed a method combining hydrothermal and ball milling to upcycle SLCO into a surface-coated LCO with a mixed-phase coating. The mixed-phase coating (Co_3O_4/Co_2AlO_4, $CoTiO_3$, and Li_3PO_4) demonstrates excellent structural stability under high voltage. Its primary function is to inhibit detrimental electrolyte interactions. Therefore, the R-LCO exhibits a high-capacity retention of 93% over 100 cycles at 1C (3.0–4.4 V). Zhang et al.[51] successfully R-LCO using a two-step direct recycling method combining hydrothermal relithiation and short-term annealing. During the hydrothermal process, doping elements (Ni and Mn) were uniformly introduced into the system, achieving Ni–Mn co-doping in the SLCO, which significantly improved its electrochemical performance. XRD results confirmed that the regenerated cathode exhibits a well-defined α-$NaFeO_2$-type layered structure (space group R-3m) with no detectable peaks of Co_3O_4 or CoO, indicating successful restoration of the layered structure. The upcycled LCO exhibits a discharge capacity of 160.23 mAh/g at 1C, with a capacity retention rate of 91.2% after 100 cycles (Figure 5.10).

Challenges of liquid-phase regeneration of LCO: Although liquid-phase repair offers advantages such as mild reaction conditions and the ability to uniformly restore the material's crystal structure, it still exhibits significant technical limitations. First, the hydrothermal reaction process demands extremely stringent control over parameters such as temperature, pressure, and reaction duration.

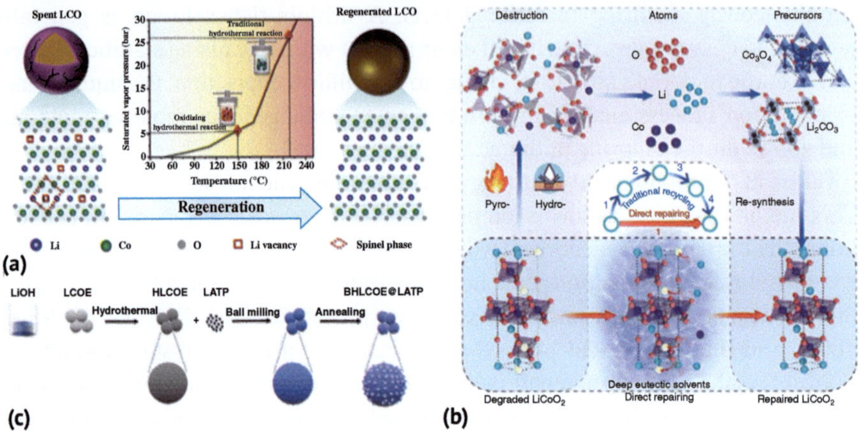

Figure 5.10 Schematic illustrations of different liquid regeneration strategies for SLCO. (a) Oxidizing hydrothermal reaction for SLCO regeneration. *Source:* [48] / with permission of Springer Nature; (b) comparison between direct DES repair and conventional regeneration routes for SLCO. *Source:* [49] / Oxford University Press / CC BY 4.0; (c) hydrothermal upcycling process for the synthesis of LCO@LATP. *Source:* [50] / with permission of Elsevier.

Even slight deviations in these conditions can easily lead to structural inhomogeneity, insufficient crystallinity, or surface impurities in the repaired material, thereby adversely affecting its electrochemical performance.[49] Secondly, the regenerated material often requires a secondary high-temperature calcination step to restore lattice activity, which increases both the processing steps and energy consumption costs.[52] Furthermore, the hydrothermal method typically employs corrosive acidic or alkaline solutions, leading to equipment corrosion.[15] This necessitates the use of high-quality, corrosion-resistant apparatus, thereby increasing the overall material cost. Considering economic factors, the hydrothermal method is primarily used for regenerating LFP materials, as NCM and LCO typically require subsequent high-temperature annealing at up to 900 °C, which significantly increases the overall process cost.[53]

Future directions of liquid-phase regeneration of LCO: Liquid-phase repair technology, exemplified by the hydrothermal method, has emerged as a research focus for regenerating high-value cathode materials owing to its mild reaction conditions and capability for precise microstructure control. The hydrothermal method facilitates lithium reinsertion via a liquid-phase environment under high temperature and pressure. However, conventional static hydrothermal reactions often suffer from significant mass transfer resistance and high activation energy for lithium-ion diffusion, resulting in low repair efficiency. Consequently, developing advanced hydrothermal strategies that enhance efficiency while reducing energy consumption represents a necessary direction for the evolution of this technology. The incorporation of additives and reactor improvements, such as enhanced mixing efficiency, can significantly optimize the regeneration outcome. Furthermore, excessive lithium salt addition leads to high concentrations of free lithium ions in the solution, increasing the burden and cost of subsequent wastewater treatment.[1] Therefore, optimizing the process pathway, reducing solvent usage, developing reusable solvents, and establishing closed-loop systems to recover lithium salts from waste solutions will enhance the economic viability and environmental benefits of hydrothermal regeneration. Additionally, severe aggregation of cathode material particles often occurs during hydrothermal regeneration, adversely affecting the repair quality. The integration of microwave and ultrasonic treatment can effectively improve particle dispersion uniformity, enhance the regeneration efficiency, and further promote the application of hydrothermal regeneration.[54, 55]

5.1.2.1.3 Liquid-phase Regeneration of NCM

Developed regions such as Europe and the United States initiated research on the regeneration of spent ternary cathode materials (SNCM) at an early stage. Research teams in these regions have achieved notable progress in liquid-phase lithium replenishment techniques. This method utilizes ionic lithium sources to directly interact with the degraded material, enabling targeted repair of lithium-deficient sites in the NCM crystal structure.[56] In contrast to solid-state sintering repair, the liquid-phase regeneration does not involve adding a stoichiometric quantity of lithium. This approach eliminates the need for predetermining the degree of lithium loss in the degraded cathode. Currently, research on liquid-phase regeneration of NCM is predominantly focused on hydrothermal methods.

To address the challenges of tedious pretreatment and the use of toxic organic solvents in conventional direct recycling of SLIBs, Gupta et al.[57] developed an integrated and scalable process for regenerating cathode black mass without organic solvents. This method combines purification and relithiation into a seamless three-step procedure: hydrothermal relithiation in alkaline solution, water washing, and annealing. By employing a lithium hydroxide solution under hydrothermal conditions (220 °C), the process simultaneously achieves relithiation and decomposition of polyvinylidene fluoride (PVDF) binder—enabling the removal of key impurities such as conductive carbon, electrolyte residues, and aluminum shreds without toxic solvents. The regenerated NCM111 exhibits a discharge capacity of 155 mAh/g with 98% capacity retention after 100 cycles. Yu et al.[58] developed a redox-mediated low-temperature hydrothermal method to address the sluggish kinetics of lithium replenishment in SNCM cathodes. By introducing low-cost mediators such as ethanol, hydrogen peroxide, and ethylene glycol, this method significantly reduces the energy barrier for relithiation. The relithiation temperature is lowered from 220 to 100 °C or below. The technology not only reduces the pressure during relithiation (from 25 to 1 bar), significantly improving operational safety, but also achieves complete regeneration of both chemically delithiated and electrochemically degraded NCM cathodes (including NCM111 and NCM622), restoring their composition, crystal structure, and electrochemical performance. Jia et al.[38] reported a regeneration strategy for spent NCM523 based on topochemical transformation. The strategy employed an initial ammonia-assisted hydrothermal pretreatment to convert the surface rock salt/spinel phases into a layered hydroxide. This transformation changed the lithium-ion transport pathways from unfavorable 2-TM

Li^+ channels to favorable 1-TM Li^+ channels, enabling rapid regeneration of spent NCM523. Furthermore, ionothermal lithiation has been applied to directly regenerate spent NCM111.[59] Using low-cost lithium salts (e.g., LiBr) and recyclable ionic liquids as solvents, international researchers successfully achieved direct regeneration of spent NCM cathodes under mild conditions at 150 °C. This approach effectively restored the crystal structure and chemical composition of NCM111 while significantly improving its electrochemical performance. Notably, the ionic liquid solvent could be efficiently recovered and reused after the reaction. By significantly reducing the temperature and pressure requirements for lithium supplementation, these advances simplify the process, enhance operational safety, and promote sustainable development in battery recycling (Figure 5.11).

Challenges of liquid-phase regeneration of NCM: In practical production, existing methods—such as high-temperature solid-state, molten salt, and hydrothermal regeneration—demonstrate certain effectiveness under laboratory conditions. However, their requirements for stringent reaction control and high purity significantly limit process efficiency and elevate costs, thereby hindering large-scale industrial adoption.[60] Specifically, current separation methods often introduce multiple impurities that compromise the purity of ternary cathode materials. For instance, binder dissolution can leave organic residues, ultrasonic treatment may

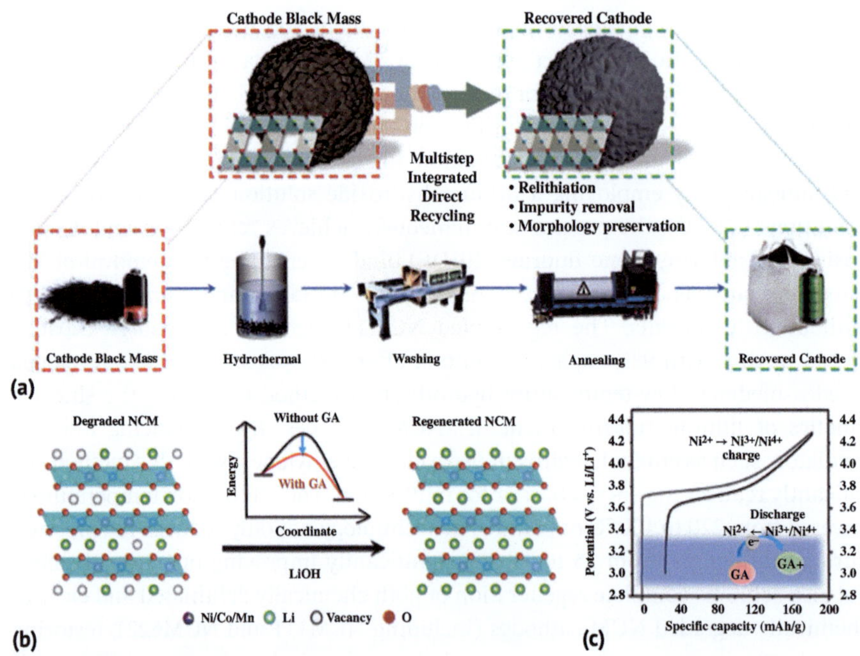

Figure 5.11 (a) Schematic diagram of the multistep integrated direct regeneration process for NCM cathode black mass. *Source:* [57] / with permission of John Wiley & Sons; (b) schematic illustration of hydrothermal relithiation-based direct regeneration process of degraded NCM; and (c) illustration of the role of green additives in the hydrothermal relithiation process. *Source:* [58] / with permission of Elsevier.

fragment aluminum foil and cause its incorporation into the cathode material, direct calcination tends to yield fluoride residues, and commercial crushing or shredding processes might introduce metallic impurities such as aluminum, copper, and iron.[13,61,62] Although aluminum and copper current collectors are the most likely impurities to remain in cathode materials after crushing, research on their impurity content and the impact on the performance of recycled materials remains relatively limited, which to some extent hinders the improvement of the purity of recycled materials. Furthermore, in the solid-state regeneration process, the absence of a liquid-phase reaction makes impurity removal more challenging. Existing repair methods are generally considered as a resynthesis process for cathode materials. However, they lack specific designs targeted at impurity removal.[21] This issue urgently needs to be addressed in the large-scale industrial application of recycled materials. On the other hand, the uncertainty in lithium supplementation is a critical factor affecting the repair efficacy. Current methods rely on compositional analysis of degraded NCM materials to determine the lithium supplementation amount. However, in practical recycling operations, factories often handle mixtures of ternary cathode materials with diverse types and varying states of health, resulting in significant compositional variations that hinder accurate lithium content determination. Insufficient lithium supplementation limits the capacity recovery of recycled materials, whereas excessive supplementation can lead to lithium salt residues after high-temperature annealing, thereby impairing electrochemical performance. Moreover, current regeneration approaches primarily target materials with high residual capacity (>50%). In contrast, for severely degraded materials (<20%) with severe structural degradation, the issues extend beyond lithium loss to include TM dissolution and complex cation disordering. As a result, simple lithium supplementation processes are ineffective in restoring the performance of these materials.[63]

Future directions of liquid-phase regeneration of NCM: Although liquid-phase regeneration demonstrates excellent performance in structural restoration, its industrial application is limited by safety risks associated with high-pressure operations and high equipment costs. Future efforts should focus on developing low-pressure hydrothermal systems or room-temperature electrochemical repair techniques to reduce costs and improve industrial feasibility. Furthermore, as this method shows limited effectiveness in repairing internal structural damage, its application to ternary cathode materials merits further investigation.

5.1.3.1 Advanced Repair Technology

5.1.3.1.1 Novel Restoration Technology for LFP

By drawing on relevant experiences from other fields, the restoration methods for LFP have been further expanded. Among these approaches, two electrochemical regeneration technologies have emerged as the primary options: one derived from the technical concept of lithium extraction from salt lakes, and the other evolved from the prelithiation strategy. Leveraging the electrochemical intercalation/deintercalation method for lithium extraction from salt lakes, an electrochemical regeneration approach has been designed—characterized by lithium deintercalation at the anode and simultaneous lithium intercalation at the cathode—to recover spent LFP cathodes from retired batteries.[64, 65] In this method, both the cathode and anode are fabricated using spent LFP from retired batteries. Driven by a potential difference, lithium ions undergo deintercalation from the LFP at the anode, followed by migration through an aqueous solution and subsequent intercalation into the spent LFP at the cathode. Ultimately, the spent LFP at the cathode is fully converted into high-purity iron phosphate, while the lithium loss in the spent LFP at the anode is effectively replenished. This novel electrochemical regeneration approach eliminates the need for an additional lithium source and is distinguished by its low energy consumption and high efficiency. In addition, drawing on the prelithiation concept,[66, 67] a functional lithium-supplementing separator was fabricated by coating a prelithiation reagent onto one side of a commercial separator. This separator was subsequently employed for the in situ electrochemical regeneration of LFP electrode sheets.[68] Replace the commercial polypropylene separator with the prepared functionalized separator, and assemble a full cell using the retired LFP electrode and commercial graphite. After the assembled full cell is activated via a single charge-discharge cycle, lithium oxalate undergoes irreversible decomposition while releasing a large quantity of lithium ions. These released lithium ions can be re-intercalated into the lithium vacancies of the retired LFP cathode, thereby achieving the objective of regeneration (Figure 5.12).

Challenges of novel repair technologies for LFP: The electrochemical regeneration technology, which draws on the concepts of lithium extraction from salt lakes, achieves material repair through the selective extraction and re-intercalation of lithium ions. However, its core challenge lies in the contradiction between electrode interface stability and reaction kinetics.[69, 70] This technology requires the construction of a specific electrolyte system to achieve efficient lithium-ion migration. However, side reactions between the passivation layer on the surface of LFP and

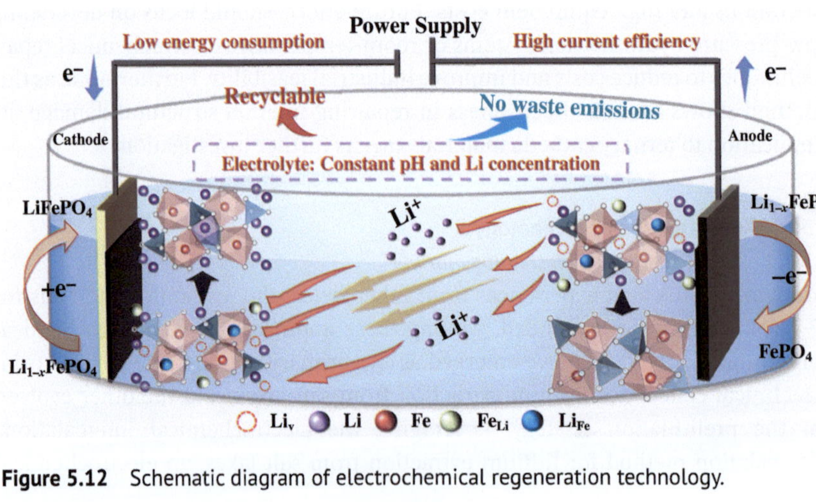

Figure 5.12 Schematic diagram of electrochemical regeneration technology.

the electrolyte may exacerbate the loss of active lithium. Particularly in long-term cycling, the interfacial impedance increases significantly, leading to a decline in repair efficiency.[71, 72] In addition, electrochemical regeneration is sensitive to the initial state of retired materials. If the materials suffer from severe structural collapse or metal impurity contamination, the lithium-ion transport channels will be blocked, making it difficult to achieve deep repair. However, although the electrochemical regeneration technology based on the prelithiation concept can restore capacity through exogenous lithium supplementation, it faces the challenge of selecting lithium supplements: existing prelithiation reagents are prone to react with air, leading to the easy introduction of by-products into the battery system.[73] Although novel technical routes are theoretically innovative, the issues of high equipment complexity and narrow process windows still need to be addressed, making it difficult to achieve low-cost continuous production in the short term.

Future directions of novel repair technologies for LFP: The electrochemical regeneration technology achieves material repair through the selective extraction and re-intercalation of lithium ions, and has advantages such as high efficiency and environmental friendliness. However, its core challenge lies in the contradiction between electrode interface stability and reaction kinetics. Side reactions between the passivation layer on the surface of LFP and the electrolyte may exacerbate the loss of active lithium, resulting in a decline in repair efficiency. In the future, it will be necessary to modify the electrolyte. By optimizing the interface structure and chemical composition, the interfacial impedance can be reduced, and the lithium-ion migration efficiency can be improved. Existing prelithiation reagents are prone to reacting with air, leading to the introduction of by-products into the battery system. In the future, it is necessary to develop novel composite lithium supplements and utilize surface modification methods to suppress air sensitivity, thereby achieving stable storage. For instance, the chemical stability of lithium supplements can be enhanced through coating technology[74] or nanostructure design,[75] while their electrochemical performance can be improved simultaneously. In addition, exploring in situ lithium supplementation technology is also an important direction. Replenishing active lithium directly via electrochemical methods can avoid introducing additional impurities and improve the consistency of the repaired materials. Electrochemical regeneration is sensitive to the initial state of retired materials. If the materials suffer from severe structural collapse or metal impurity contamination, the lithium-ion transport channels will be blocked, making it difficult to achieve deep repair.[76] In the future, it is necessary to develop high-efficiency reactors and enhance the mass transfer process to improve the lithium-ion migration rate. In addition, optimizing electrode design and electrolyte circulation systems is also an important direction, which helps to achieve low-cost and high-efficiency continuous production. In terms of equipment, the electrochemical regeneration technology features high equipment complexity and a narrow process window, making it difficult to meet the requirements of large-scale industrialization. In the future, it is necessary to develop continuous production equipment. Through modular design and intelligent control, precise regulation of process parameters can be achieved.

5.1.3.1.2 Novel Restoration Technology for LCO

Regarding the repair using novel technical routes, methods such as microwave-assisted synthesis, ultrasonic cavitation, molten salt method, eutectic solvents, and electrochemical intercalation also possess numerous advantages including low cost, environmental friendliness, and high added value. (1) Compared with the traditional hydrothermal method, the microwave hydrothermal method requires shorter reaction time and lower experimental temperature. Moreover, the powder synthesized by the microwave hydrothermal method has high purity, good crystallinity, and uniform microstructure.[55, 77] (2) Ultrasonic hydrothermal repair combines the advantages of high efficiency and clean production of ultrasonic waves and hydrothermal method, and is a method that can directly repair the crystal structure of SLCO and remove organic substances.[78] Ultrasonic cavitation generated by high-power ultrasound in the liquid phase can effectively reconstruct blocked channels and layered structures. (3) Eutectic molten salts enable rapid transport and transfer of ions in the molten state, which can be carried out at mild temperatures.[79, 80] (4) Eutectic solvents can promote the selective transport of lithium and cobalt to directly repair SLCO. Instead of dissolving LCO, they directly act as carriers for the selective supplementation of lithium and cobalt.[81] (5) Electrochemical regeneration requires the use of an external current to drive the re-intercalation of lithium ions, while reducing high-valent metals and repairing the SEI.[82] The above novel technologies integrate the synergistic advantages of multiple mechanisms, such as microwave, ultrasound, and molten salt, achieving precise structural regeneration in a green and efficient manner, and are promoting the upgrading of lithium battery recycling toward low-carbon recycling and high-value resource utilization (Figure 5.13).

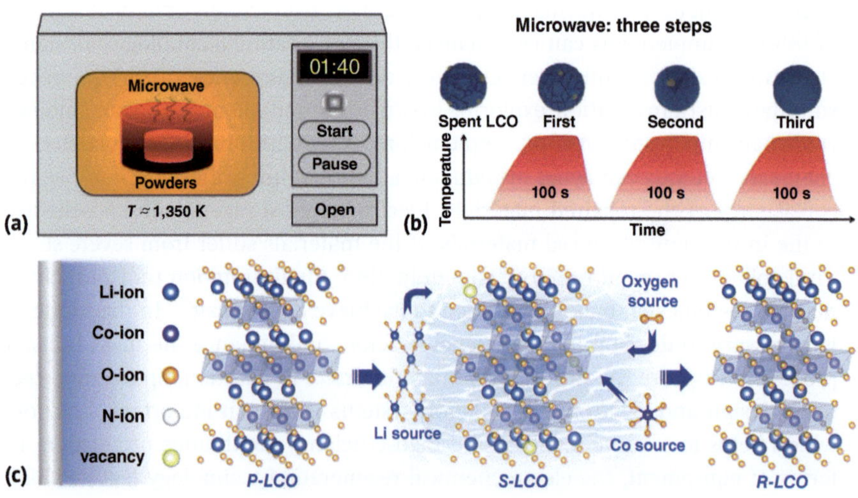

Figure 5.13 (a) Schematic diagram of microwave lithium supplementation device, (b) programmed temperature diagram during microwave reaction process, and (c) schematic diagram of structural evolution of LCO during degradation and regeneration processes.

Challenges of novel repair technologies for LCO: Regarding novel technical routes, innovative approaches such as microwave-assisted synthesis, ultrasonic cavitation, molten salt method, eutectic solvents, and electrochemical intercalation have demonstrated the potential for green and efficient performance, yet they are constrained by the bottleneck of insufficient technical maturity. (1) Although microwave-assisted hydrothermal synthesis can effectively repair material structures, the uneven distribution of electromagnetic fields easily leads to uneven heating of materials,[83] causing lattice distortion or element segregation and thus affecting material uniformity. Meanwhile, microwave equipment consumes high energy and the design of dedicated reactors is complex, making it difficult to achieve industrial production. (2) Ultrasonic cavitation technology is a process involving bubble formation and implosion. Its cavitation intensity is related to the viscosity of the solution[13]—residual binders or current collector fragments in deactivated materials will significantly weaken the cavitation effect, resulting in unstable repair performance. (3) For the molten salt method, although ion conductivity increases in the molten phase (making it more applicable to materials with different degrees of deactivation and thus ensuring regeneration performance), finding a suitable molten salt system without introducing external impurity ions is a key challenge for the molten salt roasting method. Different lithium salt systems may be required for the regeneration of different cathode materials. Additionally, the subsequent recovery of molten salts is difficult, and materials usually need secondary sintering after washing, posing environmental and economic risks.[84] (4) Although the eutectic solvent method is environmentally friendly, it has high requirements for solvent design and material compatibility. Some deep eutectic solvents have poor dissolution selectivity for LCO, which easily causes the dissolution of TMs.[85] (5) The electrochemical intercalation method can precisely regulate the re-intercalation of lithium ions, but it requires the construction of highly stable electrode and electrolyte systems. Moreover, it cannot solve the problem of structural damage at present and often requires additional sintering processes.

Future directions of novel repair technologies for LCO: Although novel technical routes (such as microwave-assisted synthesis, ultrasonic cavitation, molten salt method, eutectic solvents, and electrochemical intercalation method) have demonstrated significant potential in terms of green efficiency and precise regulation, their industrialization process is constrained by core issues including insufficient technical maturity, high equipment costs, and poor process adaptability. Thus, there is an urgent need to address the following key requirements: (1) For microwave-assisted hydrothermal synthesis technology, it is necessary to solve the problems of electromagnetic field uniformity regulation and equipment cost reduction with efficiency improvement. The microwave method can rapidly repair materials through dielectric heating, but the uneven distribution of the electromagnetic field leads to local temperature differences, resulting in insufficient material uniformity. In addition, microwave reactors are costly and have a low processing capacity for waste materials per batch, making it difficult to meet the demand for large-scale production. In the future, it is required to develop multi-module source radiation systems and implement real-time temperature control settings to improve heating uniformity, while reducing the cost of reactors. (2) For ultrasonic cavitation technology, it is necessary to enhance the cavitation effect and improve process stability. Ultrasound relies on high-frequency sound waves to generate cavitation bubbles for accelerating lithium mass transfer. However, residual binders or metal fragments in materials will significantly weaken the cavitation effect, and energy attenuation is prone to occur during continuous processing, leading to unstable repair performance. Therefore, when applying this technology, strict impurity control is required to ensure the purity of the original materials, avoid subsequent impacts, and create a favorable liquid-phase mass transfer environment. In addition, considering the high economic cost caused by long-term ultrasound and the different ultrasonic frequencies required for different materials, pulsed ultrasound can be used to adjust the power level to adapt to the regeneration of different materials and reduce energy consumption. (3) The technical requirements for the molten salt method stem from the optimization of molten salt system adaptability and the development of a closed-loop recycling process. The molten salt method uses low-melting-point mixed salts as ion-conducting media, enabling rapid intercalation of lithium ions and achieving significant regeneration effects on materials with severe structural collapse. However, systematic research on the design of eutectic media is currently lacking, making component design challenging. Molecular dynamics simulations and first-principles calculations can be employed, supplemented by studies on experimental parameters under different temperatures, times, and reactant ratios, to repair various degraded cathode materials. In the future, it will be necessary to construct a molten salt component database and use machine learning to screen molten salt systems with low eutectic points, high lithium solubility, and environmental compatibility.[86] (4) The eutectic solvent method needs to overcome the challenges of solvent selectivity design and inhibition of TM dissolution. Eutectic solvents (such as lithium chloride-urea systems), with their green properties, low cost, and high selective transport of lithium/cobalt, can achieve mild repair at low temperatures (80 °C) and atmospheric pressure. The selected eutectic solvents must have low solubility for LCO to reduce the dissolution rate of metal ions during the repair process.[49] Additionally, the adaptability of solvents to different cathode materials (e.g., NCM and LFP) varies significantly. In the future, molecular dynamics simulations should be used to reveal the mechanism of action of solvents on waste materials, and additives (such as corrosion inhibitors and reducing agents) should be added to inhibit metal dissolution. (5) The core technical requirements of the electrochemical intercalation method lie in the innovation of electrode/electrolyte systems and the synergistic optimization of structural repair. By controlling the electrolyte, current, and voltage, electrochemical methods exhibit high selectivity for depositing different ions, enabling materials to undergo lithium supplementation and doping simultaneously. However, existing electrolytes are prone to oxidative decomposition under high voltages, and electrochemical repair cannot effectively repair deep cracks and defects in materials. In the future, it will be necessary to develop highly stable solid electrolytes that can withstand high voltages and inhibit side reactions. Furthermore, real-time monitoring of the regeneration process (using in situ X-ray diffraction [XRD], Raman, FTIR, TEM technologies, etc.) is required to better understand the regeneration mechanism and synchronously repair particle microcracks during electrochemical intercalation.

5.1.3.1.3 Novel Repair Technologies for Ternary Materials

In recent years, significant progress has been made in the research on regenerating LIB cathode materials using electrochemical methods. The core of this technology lies in restoring material performance through ion migration and structural reconstruction driven by an electric field, which avoids the complex leaching-resynthesis process in traditional hydrometallurgy. This field mainly focuses on three technical directions: direct electrochemical lithium supplementation, in situ structural repair, and interface engineering optimization.[87, 88] Its innovation is reflected in the targeted repair of intrinsic material defects and the low-energy consumption characteristics of the resource utilization process.

In terms of direct electrochemical lithium supplementation, researchers have constructed lithium source-assisted electrode systems,[82] using constant potential or constant current polarization to promote the directional intercalation of lithium ions into lithium-deficient cathodes. For example, metallic lithium or lithium-rich compounds are used as counter electrodes, and the lithium content recovery of layered materials such as LCO and ternary cathode materials is achieved through electrochemical cycling in organic electrolyte systems. Electrochemical lithium pump technology realizes the recovery of the capacity of ternary cathode materials to their initial values by designing a three-electrode system with a lithium-selective permeable membrane, and its cycle stability is superior to that of commercial new products. The breakthrough of such methods lies in the development of an SEI layer with high ionic conductivity, which effectively inhibits side reactions and improves lithium transport efficiency, enabling the lithium supplementation depth to reach more than 95% of the material's theoretical capacity. In situ structural repair technology achieves self-healing of crystal defects through electrochemical field regulation. Aiming at problems such as cation mixing, oxygen vacancies, and microcracks generated during cycling, researchers have found that constant voltage holding in a specific potential range can induce the rearrangement of TM ions. Pulsed high-potential treatment of LFP can eliminate crack propagation caused by lattice stress. The kinetic mechanism of this process involves electric field-driven vacancy migration and lattice relaxation, and the reverse transformation from spinel phase to layered phase has been confirmed by in situ XRD (Figure 5.14).

Analysis of technical shortcomings and gaps: Technical demand analysis: for novel technology-based repair routes such as electrochemical repair, the current research focus is mainly on LCO, while the repair of retired ternary cathode materials remains immature. Additionally, this repair method has limited capability in restoring crystal structures. In the future, efforts should be made to explore repair processes for multi-series ternary materials, addressing issues such as the complex

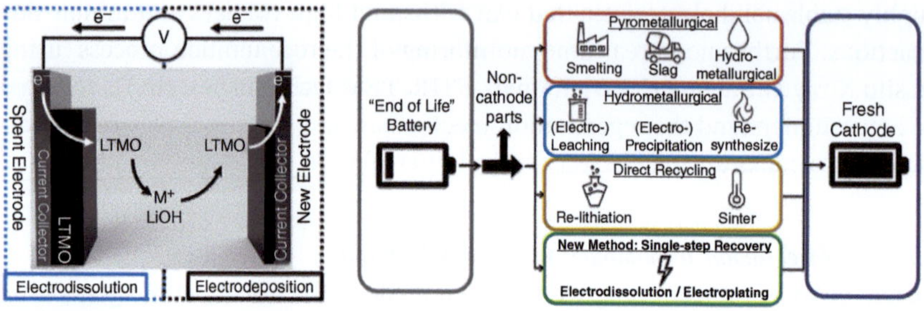

(a) (b)

Figure 5.14 (a) Simplified schematic diagram of one-step electrodissolution and electrochemical regeneration for lithium transition metal oxide (LTMO) battery cathode materials. (b) Simplified comparison diagram of common battery recycling methods and one-step electrochemical recycling method.

crystal phase structures and diverse doping elements of different ternary material series. Meanwhile, combined with technologies like topological transformation and ion implantation, research should be conducted on how to repair rock-salt phase defects and lithium loss in ternary materials through electrochemical means.

5.2 Direct Regeneration of Spent Graphite Anode

As a core component of LIBs, graphite anode materials exert a direct impact on the energy density, cycle life, and safety performance of batteries. During the battery operation process, however, graphite anode materials inevitably undergo issues such as structural degradation, surface side reactions, and impurity accumulation—all of which result in the deterioration of their electrochemical performance. Consequently, the repair technology for graphite anode materials has emerged as a crucial research avenue for enhancing battery performance, prolonging service life, and reducing production costs. Currently, the repair technologies for graphite anode materials are primarily categorized into two types: low-temperature repair technology and high-temperature repair technology. Each exhibits distinct characteristics with respect to repair mechanism, application scope, and efficacy, while also possessing certain inherent limitations. This article will elaborate on the current status, advantages, limitations, and future prospects of low-temperature repair technology and high-temperature repair technology in detail.

5.2.1 Low-temperature Repair Technology

Low-temperature repair technology refers to the treatment of graphite anode materials at relatively low temperatures (typically below 500 °C), aiming to remove surface impurities, repair surface defects, or enhance surface properties. This technology mainly includes methods such as low-temperature impurity removal, molten salt method, and surface coating. Due to their simple operation, low energy consumption, and low equipment requirements, these techniques have gained widespread attention. However, their repair effect is primarily limited to the graphite surface and cannot address internal crystal structure defects of graphite.

1. **Low-temperature impurity removal repair technology:** Low-temperature impurity removal technology mainly relies on chemical or physical methods to eliminate impurities adsorbed on the surface of graphite, such as electrolyte decomposition products, metal ions, and organic contaminants. Common approaches include acid washing, alkali washing, and solvent cleaning. Acid washing involves treating graphite materials with acidic solutions (hydrochloric acid, sulfuric acid, or nitric acid) to remove surface metal oxides and certain organic impurities. This method can effectively restore the electrochemical activity of graphite; however, it may also cause structural damage, especially under conditions of high acid concentration or prolonged treatment. Furthermore, thorough rinsing is required after acid washing to remove residual acids, as these may negatively affect battery performance. Alkali washing is primarily used to eliminate acidic impurities from the graphite surface, such as acidic by-products generated from electrolyte decomposition. Typical alkaline solutions include sodium hydroxide or potassium hydroxide. Although alkali washing has limited effectiveness in restoring graphite performance and may introduce new impurities, it is often combined with other regeneration methods for better results. Solvent cleaning utilizes organic solvents such as ethanol, acetone, or N-methyl-2-pyrrolidone (NMP) to dissolve organic contaminants on the graphite surface, thereby recovering its electrochemical activity. This method offers the advantages of simple operation and minimal structural damage to graphite, though its effectiveness against stubborn impurities remains limited. A research team from Central South University developed a two-step calcination process to recover spent graphite.[89] First, the material was treated at 400 °C for 1 hour under argon to decompose binders and separate the copper foil. Then, oxidation was performed at 500 °C in air for 1 hour to convert residual metallic copper into copper oxide and remove electrolyte by-products. The leaching efficiencies of lithium, copper, and aluminum were all close to 100%. After acid leaching, the graphite was washed with ethanol to neutrality and dried, yielding high-purity graphite that could be reused in LIBs. The regenerated graphite (RG) underwent acid exfoliation and pyrolytic reconstruction, forming thinner layered structures. Although its specific surface area decreased, the average pore size increased significantly, improving lithium-ion diffusion efficiency. Due to the enlarged interlayer spacing and enhanced surface adsorption induced by the low-temperature impurity-removal treatment, the material exhibited a capacity retention of 97.9% and nearly 100% coulombic efficiency after 100 charge-discharge cycles. The advantages of low-temperature impurity-removal technology lie in its simplicity and low energy consumption. However, its main limitations include the inability to repair internal crystal defects and insufficient removal of certain stubborn impurities. Moreover, chemical treatments may introduce new contaminants or cause damage to the

graphite structure, making strict process control essential. A Chinese Academy of Sciences research team proposed a low-temperature deep impurity-removal and epitaxial-repair strategy based on KOH-NaOH composite alkali etching for the green regeneration of spent graphite anodes.[90] Through composite-alkali roasting at 180–300 °C, acid-insoluble impurities (e.g., Al and Fe compounds) were effectively removed, reducing impurity concentrations from several hundred ppm to below 10 ppm. Subsequent acid washing and filtration further purified the graphite. Finally, pitch coating followed by 1,100 °C heat treatment enabled surface reconstruction and densification. The transformation of the coating layer from amorphous carbon to graphitic carbon at high temperatures, and its epitaxial bonding with the graphite substrate, reconstructed efficient Li^+ transport pathways. The RG exhibited electrochemical performance comparable to commercial graphite, and full cells assembled with commercial LFP maintained 85.8% of their capacity after 500 cycles at 1.0C (Figure 5.15).

2. **Low-temperature molten salt repair technology:** Molten salt repair technology is a low-temperature treatment method that utilizes the chemical reactivity of molten salts to repair graphite surfaces. By selecting an appropriate molten salt system, oxidation–reduction reactions can occur on the graphite surface at relatively low temperatures, thereby removing surface impurities and repairing certain structural defects. Common molten salt systems include chlorides, fluorides, and nitrates. The choice of molten salt has a significant impact on the repair performance—for example, chloride-based molten salts can effectively remove metal oxides from the graphite surface, whereas fluoride-based molten salts exhibit superior removal efficiency for organic impurities.

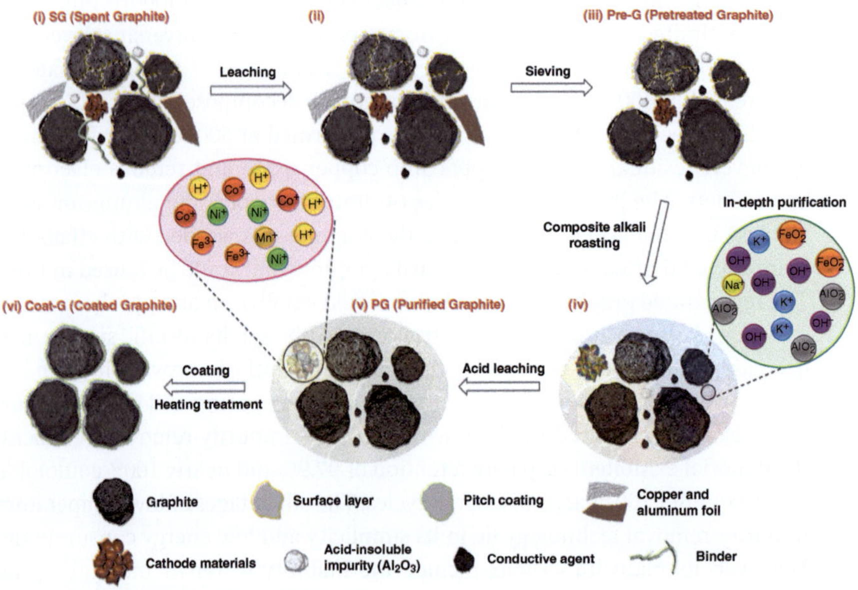

Figure 5.15 Schematic illustration of the purification and regeneration process of SG. *Source:* [90] / with permission of American Chemical Society.

The molten salt method achieves impurity removal and defect repair through the chemical interaction between molten salts and the graphite surface. A research team at Central South University successfully employed sodium-based salts (NaF, NaCl, and NaI) to realize efficient low-temperature repair of waste graphite.[91] Notably, the introduction of molten salts improved the crystallinity and isotropy of the repaired graphite, effectively mitigating the interlayer anisotropy often intensified by deep purification in hydrometallurgical processes, thereby enhancing the electrochemical performance. As a lithium storage anode, the optimized sample demonstrated a capacity of 352 mAh/g after 200 cycles at 1.0C, and maintained approximately 175 mAh/g at 5.0C (Figure 5.16).

3. **Low-temperature surface coating repair technology:** Surface coating technology improves the electrochemical performance of graphite by coating its surface with a protective layer of materials such as carbon, metal oxides, or polymers. The coating layer suppresses side reactions between the electrolyte and graphite surface, reduces the nonuniform growth of the SEI film, and thereby enhances the cycling stability and rate capability of graphite. Amorphous carbon, when used as a surface coating material for waste graphite, can compensate for defects caused by repeated cycling, repair surface structural damage, improve the degree of graphitization, and inhibit electrolyte decomposition. Moreover, the isotropic nature of the carbon source facilitates random transport of lithium ions and electrons within graphite particles, reduces irreversible capacity loss, and consequently enhances the rate performance of RG.[92] A research team at Central South University proposed a carbon-chain regulation coating strategy for repairing graphite anodes,[93] which effectively corrected pore defects on the surface of waste graphite anodes and enhanced the reactive sites within RG anodes. This approach reduced SEI film thickness while optimizing SEI stability, decreased irreversible Li^+ consumption, and improved the cycling stability of LIBs. The Coulombic efficiency significantly recovered from 14 to 86.89%, and after 100 cycles, the capacity remained around 384.9 mAh/g. At a 2C rate, the specific capacity was maintained at approximately 200 mAh/g, meeting the standards of commercial graphite materials. Additionally, a research team at the Chinese Academy of Sciences proposed

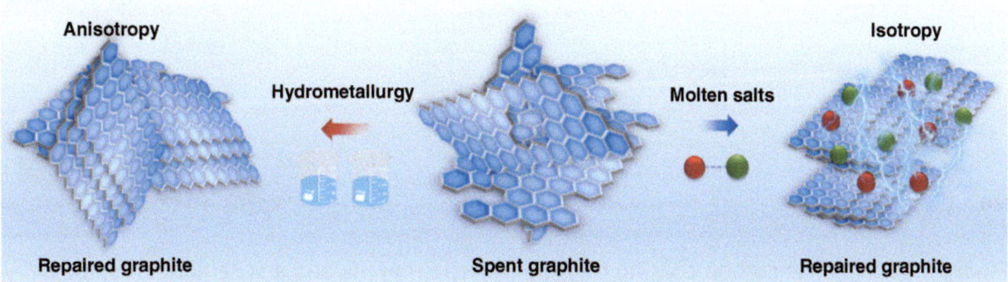

Figure 5.16 The mechanism of repairing spent graphite. *Source:* [91] / with permission of Elsevier.

a low-temperature polymerization strategy using methyl methacrylate (MMA) to coat purified waste graphite with a PMMA-based artificial SEI layer.[94] This process forms a bilayer SEI structure composed of an inner PMMA layer and an outer thin natural SEI layer. The PMMA layer not only promotes the formation of a uniform natural SEI but also effectively suppresses electrolyte decomposition and mitigates lithium loss caused by continuous SEI growth during cycling. The RG assembled into full cells exhibited a discharge capacity of 149 mAh/g after 500 cycles at 1C, with a capacity retention of 86.7% and significantly improved initial Coulombic efficiency, providing a cost-effective and energy-efficient approach for waste graphite regeneration (Figure 5.17).

Although low-temperature repair technology has achieved certain success in surface modification and performance enhancement of graphite anode materials, its main limitation lies in the inability to address crystal structure defects within graphite. During battery cycling, graphite may experience interlayer delamination, lattice distortion, and microcrack formation, which significantly deteriorate its lithium-ion intercalation/deintercalation capability and structural stability. Due to the relatively low reaction temperature, low-temperature repair methods are generally ineffective in repairing the crystalline structure of graphite, and their effects are mainly confined to local surface improvements rather than fundamental structural restoration.

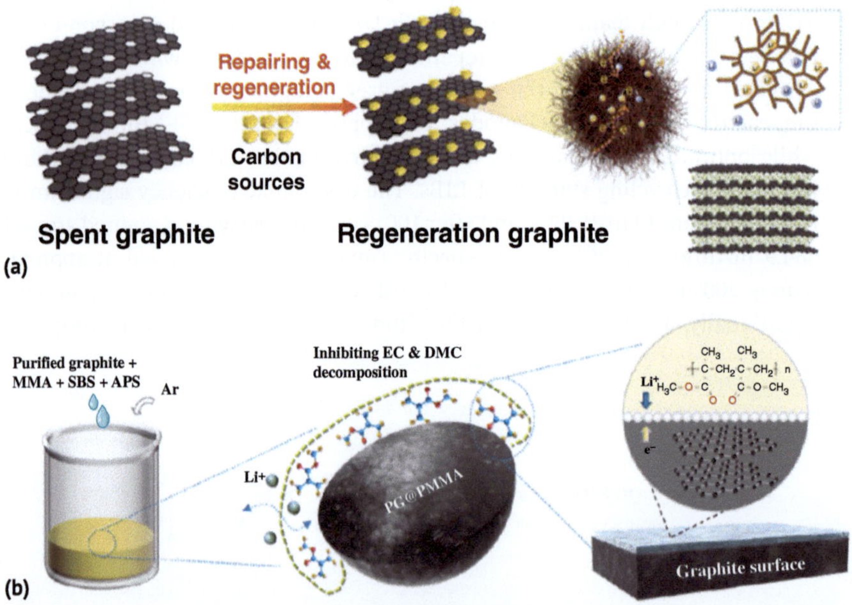

Figure 5.17 (a) The carbon coating method for repairing SG, and a schematic of electron conduction and Li^+ diffusion in graphite particles. *Source:* Adapted from [95].
(b) Schematic of the preparation and functional mechanism of PG@PMMA through emulsion polymerization. *Source:* [94] / with permission of Elsevier.

Challenges of low-temperature repair technology: Among the existing graphite anode repair methods, low-temperature repair technologies have attracted extensive attention due to their simplicity, low energy consumption, and minimal equipment requirements. However, their major drawback is the limited ability to repair internal crystal structure defects of graphite. These techniques mainly include low-temperature impurity removal, molten salt treatment, and surface coating, which primarily focus on surface purification or performance enhancement but have limited effectiveness in repairing internal defects such as interlayer delamination, lattice distortion, and microcracks. Low-temperature impurity removal methods—such as acid washing, alkali washing, and solvent cleaning—primarily eliminate surface-adsorbed impurities (e.g., electrolyte decomposition products, metal ions, and organic contaminants) through chemical or physical means. Although these methods can effectively restore the electrochemical activity of the graphite surface, their influence remains confined to the surface region, making them ineffective for repairing internal crystal defects. Similarly, molten salt treatment utilizes the chemical reactivity of molten salts to repair the graphite surface. While it can remove surface impurities and heal partial surface defects at relatively low temperatures, its capability to repair internal structural defects remains limited. Surface coating technology, on the other hand, improves electrochemical performance by depositing protective layers (e.g., carbon, metal oxides, or polymers) on the graphite surface. However, its function primarily focuses on suppressing side reactions between the electrolyte and graphite, rather than fundamentally restoring the internal crystal structure. In summary, although low-temperature repair technologies have proven effective in surface modification and performance enhancement, their limited ability to restore internal crystal integrity restricts their overall contribution to battery performance improvement. Future advancements should focus on integrating high-temperature repair processes or developing novel synergistic methods to achieve both surface and internal defect repair in graphite materials.

Future directions of low-temperature repair technology: The future development of anode repair technologies will emphasize achieving precise internal defect repair under controlled energy consumption, thereby comprehensively enhancing the performance of LIBs. Although high-temperature thermal treatment can effectively repair internal defects, its high energy demand, complex process, and poor localization control hinder large-scale application. Therefore, there is a strong need to develop low-energy, high-efficiency repair technologies—such as low-temperature plasma repair and microwave-assisted repair—which can penetrate graphite structures at lower temperatures to heal interlayer delamination, microcracks, and lattice distortions while minimizing energy consumption. Furthermore, the combination of in situ characterization techniques and artificial intelligence (AI) algorithms will enable precise localization and repair of graphite defects, providing scientific guidance for process optimization. For instance, high-resolution transmission electron microscopy (HRTEM) and Raman spectroscopy can be used to monitor defect evolution in real time, while machine learning can optimize repair parameters to improve efficiency. In terms of multifunctional composite repair, future studies should focus on synergistic techniques that integrate surface coating with internal repair or combine chemical and physical repair mechanisms to address both surface and subsurface defects. For example, a chemical vapor deposition (CVD)-based surface coating can be combined with plasma repair to achieve dual-level restoration of graphite structure. Finally, strengthening fundamental research and technology transfer—such as establishing a comprehensive graphite defect repair database and promoting industry-academia collaboration—will accelerate the industrialization of repair technologies, ultimately meeting the growing market demand for high-performance LIBs and realizing precise repair of internal graphite defects.

5.2.2 High-temperature Graphitization Repair Technology

The high-temperature graphitization repair technique for graphite anodes represents an advanced regeneration approach designed to restore the electrochemical performance and structural integrity of spent graphite anodes from LIBs under controlled high-temperature conditions. The fundamental principle involves the thermal elimination of residual binders, conductive additives, metallic impurities, and electrolyte-derived by-products on the graphite surface and between the graphitic layers, while simultaneously repairing structural defects such as microcrystalline distortion and pore collapse induced by repeated lithium-ion intercalation and deintercalation. Through high-temperature treatment, typically exceeding 1,300 K, the graphite lattice undergoes rearrangement, thereby enhancing the degree of graphitization, crystallinity, and structural stability. The high-temperature environment thermally activates atomic rearrangement within the graphite lattice, effectively mitigating interlayer delamination and lattice distortions, which leads to improved electrochemical properties and mechanical robustness.[92] Current research efforts on the high-temperature regeneration of graphite anodes primarily focus on optimizing key process parameters, including temperature range, atmospheric conditions (e.g., inert gas protection), and thermal duration, and elucidating the structural evolution mechanisms that govern electrochemical performance enhancement. Recently, the conventional single-step calcination process has evolved into multistage synergistic treatment systems. Among these, flash Joule heating (FJH) has emerged as a particularly promising technique. FJH utilizes the instantaneous discharge of a capacitor, leveraging the transient variation in electrical resistance of the reactants to generate extremely high current densities at relatively low voltages. This process produces ultrafast temperature spikes (~2,850 K) within milliseconds, facilitating the reconstruction of degraded graphite anodes and the rapid removal of surface contaminants. Compared with conventional furnace-based thermal treatments characterized by slow heating rates and long residence times, the FJH process significantly reduces regeneration costs, greenhouse gas emissions, water consumption, and overall energy input. The research group led by Liangbing Hu at the University of Maryland developed a continuous high-temperature direct-recycling process for impurity-laden degraded graphite.[96] Employing an inclined carbon heater operating at approximately 2,000 K, the waste graphite descends through the heating zone, completing the rejuvenation process within 0.1 seconds, thereby enabling scalable, high-throughput regeneration. This approach effectively eliminates impurities such as SEI components and polymeric binders, restores the layered graphite architecture, enhances the graphitization degree, and slightly enlarges the (002) interlayer spacing, beneficial for Li^+ intercalation/deintercalation kinetics. The RG paired with a lithium counter electrode exhibits a reversible capacity of 320 mAh/g at 1.0C and maintains 96% capacity retention over 500 cycles. Similarly, the research team led by Shengming Xu at Tsinghua University systematically investigated graphite anode degradation mechanisms and developed an FJH-based recycling route enabling efficient repair and performance enhancement across diverse waste graphite sources,[97] including manufacturing scrap, dismantled electrodes, and hydrometallurgical residues. The process involves a two-stage vacuum treatment: an initial 36 V preheating stage followed by a 180 V pulsed current stage, during which transient high temperatures (~3,000 °C) are achieved within milliseconds. The intense Joule heating rapidly decomposes SEI layers, binders, and embedded molecular residues while simultaneously repairing lattice defects and restoring graphitic order. The entire procedure completes in under one second—two to three orders of magnitude faster than conventional thermal sintering. The RG demonstrates outstanding electrochemical performance, with a specific capacity of 358 mAh/g at 0.1C, an initial Coulombic efficiency of 87.8%, and a capacity retention of 94.83% after 100 cycles. Morphological and structural characterizations (SEM, XRD, and Raman spectroscopy) confirm the effective removal of surface impurities, restoration of a uniform interlayer spacing (~0.335 nm), and substantial improvement in carbon atom ordering. From an economic and environmental standpoint, the FJH process reduces energy consumption by 82%, greenhouse gas emissions by 95.8%, and water usage to zero, resulting in a per-kilogram profit of approximately 57.75 RMB, underscoring its industrial scalability. Density functional theory (DFT) simulations further elucidate that the ultrahigh current density in FJH induces ordered layer realignment, providing both mechanistic insight and theoretical support for the efficient regeneration of spent graphite. Furthermore, the research team led by Yunhui Huang at Huazhong University of Science and Technology proposed a rapid high-temperature repair methodology for degraded graphite anodes,[98] achieving efficient regeneration via transient Joule heating at approximately 1,900 K for 150 ms. This process rapidly reconstructs the SEI layer, transforming the originally loose, organic-rich thick SEI (~100 nm) into a compact, inorganic-dominated thin layer (~10–30 nm) composed primarily of LiF, Li_2CO_3, and Li_2O. The resulting layer exhibits a 26-fold increase in Young's modulus, significantly enhancing interfacial mechanical stability. Meanwhile, the short-duration thermal exposure preserves partial residual active lithium without damaging the graphite lattice, imparting the regenerated material with intrinsic prelithiation characteristics. The rejuvenated graphite electrodes deliver exceptional electrochemical performance, including an initial Coulombic efficiency of 104.7%, superior cycling stability, and excellent rate capability surpassing commercial graphite. This rapid high-temperature process not only reduces energy consumption (~79%) and carbon emissions (~91%) but also enables high-value direct upcycling of "waste graphite to high-performance prelithiated graphite." Overall, this approach demonstrates remarkable economic and environmental benefits and provides a sustainable technological pathway for the rapid and efficient high-temperature rejuvenation of spent graphite anodes (Figure 5.18).

Challenges of high-temperature technology: The high-temperature graphitization repair technology still faces several technical challenges in practical applications, with high energy consumption being one of the core issues that severely restricts its large-scale deployment and economic feasibility. Firstly, the high energy consumption of graphite repair primarily originates from its complex process flow. The graphite regeneration process typically involves multiple stages, including physical separation, chemical purification, and high-temperature thermal treatment. For example, during high-temperature regeneration, graphite materials must undergo thermal treatment at temperatures typically exceeding 1,000 °C to remove surface impurities, repair structural defects, and restore

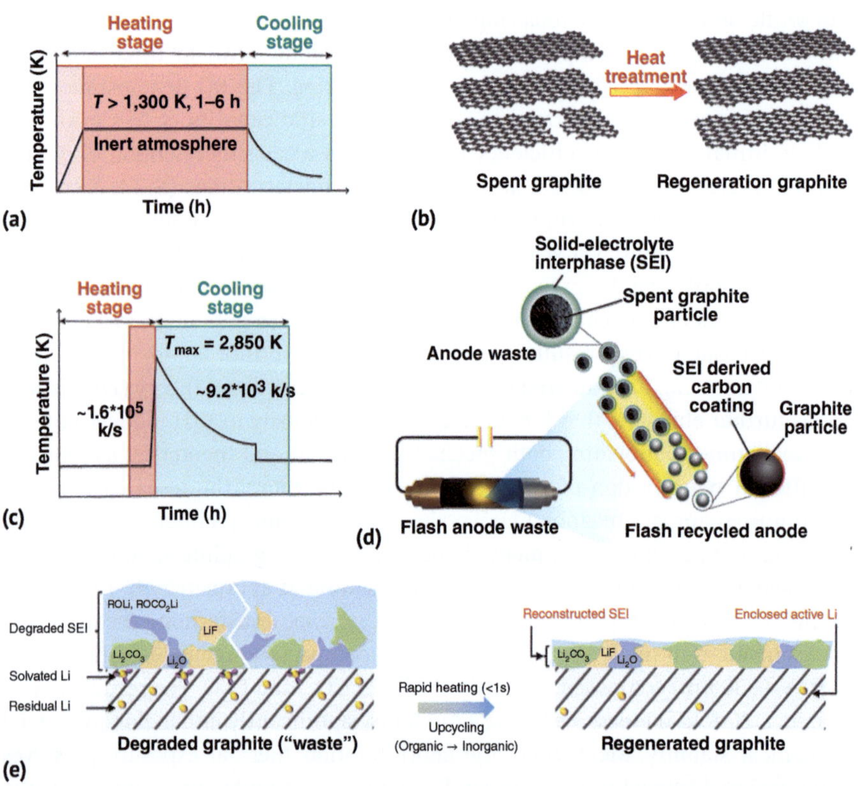

Figure 5.18 (a, c) The general procedure and real-time temperature curves of the heat treating and flash recycling methods, respectively. (b) Comparison diagram of SG before and after high-temperature treatment. (d) Schematic diagram of SG regeneration based on FJH method. *Source:* [92] / with permission of Elsevier. (e) Schematic diagram showing the transformation of degraded graphite with loose SEI into RG with compact one. *Source:* [99] / with permission of John Wiley & Sons.

their electrochemical activity. This process not only requires substantial thermal energy input but also demands long-duration temperature maintenance, leading to significantly increased energy consumption. In addition, high-temperature treatment carries risks such as graphite oxidation or structural degradation, further increasing both process complexity and energy usage. Secondly, the high energy demand of graphite repair is closely related to the stringent equipment requirements. To achieve efficient repair, the regeneration equipment must provide precise temperature control, uniform thermal field distribution, and high energy utilization efficiency. However, most current systems exhibit poor thermal efficiency and significant heat loss, resulting in persistently high energy consumption. Moreover, equipment operation and maintenance costs remain substantial, further elevating the overall energy footprint. Additionally, the high energy consumption of graphite repair is linked to the stringent purity requirements of the feedstock. During the repair process, graphite materials typically undergo multiple washing and purification steps to remove metallic impurities and organic residues. These processes not only consume large quantities of water and chemical reagents but also incur additional energy expenditure. Particularly when treating heavily degraded or complex graphite materials, the repair process becomes more energy-intensive, undermining its economic viability and sustainability.

Future directions of high-temperature technology: Optimizing the balance between energy efficiency and repair effectiveness is key for future advancements. Reducing the repair temperature and duration, as well as developing continuous regeneration processes, will be crucial directions. For instance, microwave-assisted repair technology can achieve efficient regeneration within a short time while significantly reducing energy input, and continuous repair systems can enhance production efficiency and lower operational costs. Furthermore, promoting environmentally friendly repair technologies represents an important future trend, such as employing non-toxic, pollution-free reagents, and enabling by-product recycling to minimize environmental impact. Through these advancements, the future development of graphite anode repair technologies is expected to achieve low energy consumption while simultaneously enhancing efficiency, thereby advancing LIB technology and contributing to sustainable growth in the new energy sector.

Final summary of graphite repair: The future development of waste graphite regeneration technologies will focus on overcoming existing technical bottlenecks and achieving synergistic restoration of both surface and internal defects. On the one hand, low-temperature repair technologies should be expanded toward the remediation of deep structural defects. By developing novel low-temperature plasma and microwave-assisted repair methods, it is possible to maintain low energy consumption while precisely repairing interlayer delamination and microcracks, thus avoiding the energy burden associated with conventional high-temperature treatments. On the other hand, high-temperature repair technologies must emphasize energy efficiency optimization by exploring innovative processes such as transient high-temperature treatment and gradient heating to shorten processing time and reduce heat loss, while employing inert gas atmospheres to mitigate oxidation risks. Key breakthroughs include promoting hybrid process integration—such as "low-temperature surface purification and high-temperature structural repair" or "molten salt pretreatment and pulsed Joule heating"—to simultaneously remove surface impurities and reconstruct the crystal lattice. Additionally, the integration of intelligent control strategies combining in situ characterization with AI algorithms can enable real-time defect monitoring and dynamic optimization of repair parameters, achieving targeted precision repair. The transition toward green process upgrading will also be essential, involving the development of pollution-free chemical systems, by-product recovery mechanisms, and low-carbon, circular repair workflows. Moreover, realizing continuous production through highly integrated repair lines will enhance industrial scalability and economic efficiency. Through the synergistic optimization of cross-scale defect healing and energy efficiency, graphite regeneration technologies will continue to evolve toward low-energy, high-efficiency, and value-added directions, providing a robust materials foundation for the sustainable advancement of the LIB industry.

5.3 Final Chapter Summary

Direct regeneration of LFP: The future development of LFP battery repair technologies requires systematic breakthroughs in process optimization, material innovation, and equipment upgrades. Solid-state repair technologies need to improve process control precision and material compatibility; liquid-phase repair technologies should focus on developing green processes and efficient wastewater treatment solutions; electrochemical regeneration technologies need to overcome challenges in interface engineering and lithium supplementation. Through cross-disciplinary collaborative innovation, a full-chain technology system and recycling industry chain can be established. LFP battery repair technologies are expected to achieve low-cost, high-efficiency, and large-scale industrial applications, promoting the sustainable development of the lithium battery industry.

Direct regeneration of LCO: Solid-state and liquid-phase repair technologies are crucial methods for regenerating LIB cathode materials. Solid-state methods are limited by low solid-to-solid contact efficiency, which results in slow and uneven lithium ion diffusion. Therefore, methods such as high-energy ball milling, crushing, mechanochemical activation, and rapid Joule heating are key choices to reduce calcination temperature and time and improve mixing uniformity. Liquid-phase repair technologies face challenges such as high mass transfer resistance, high lithium activation energy, and excessive lithium salts, which complicate subsequent wastewater treatment. Additive optimization, equipment improvements, and the use of microwave and ultrasound combined treatments can enhance repair efficiency and reduce energy consumption. Meanwhile, new technological approaches, including microwave-assisted, ultrasound cavitation, molten salt methods, eutectic solvents, and electrochemical embedding, show potential in green efficiency and precise control, although industrialization still faces challenges like equipment costs, technological maturity, and process compatibility. The demand analysis of these technologies should focus on precise control and industrialization, breaking through the bottlenecks of efficiency, cost, and environmental impact to advance lithium battery regeneration technology toward high-efficiency, low-carbon solutions.

Direct regeneration of NCM: An important gap in current research is the lack of in-depth investigation into the failure mechanisms of NCM cathode materials. Future research should move beyond simply relying on battery performance data and basic elemental analysis, instead focusing on integrating theoretical calculations. Researchers need to uncover the potential mechanisms during the cycling process and validate theoretical predictions using in situ characterization techniques. Given the limitations of individual in situ methods, advancing high-throughput in situ studies is imperative. Before implementing repairs, it is essential to conduct a comprehensive analysis of the failure mechanisms of cathode materials to select the optimal repair approach.

Specifically, for materials with minor structural damage and quantifiable lithium deficiency, high-temperature solid-state methods are recommended, as they create favorable conditions for lattice rearrangement and stress relief, even though their lithium-ion transport efficiency at the solid-phase reaction interface is relatively low. For materials with undetermined lithium deficiency but minimal structural damage, hydrothermal methods are preferred, due to their high reaction efficiency, although they have limited effectiveness in repairing internal structural damage. For materials with severe structural damage, molten salt methods are the best choice due to their excellent ion diffusion efficiency and thorough structural recovery.

It is worth noting that while electrochemical repair methods are environmentally friendly, their application in NCM cathode materials is still underexplored and requires further investigation. Since direct regeneration of cathode materials is a systematic process, researchers should construct a comprehensive system model to address key issues like battery design standardization, fault evaluation standardization, and market "identifiability" of current batteries. This model should aim to integrate current research outcomes into a system with practical application value.

It is crucial to emphasize that almost all cathode material repair methods require composition analysis of degraded materials to determine the amount of lithium to supplement before re-lithiation. While this approach seems reasonable, it is practically unfeasible in production. Recycling facilities crush and sort large quantities of waste lithium batteries, resulting in black blocks that are mixtures of different battery cathode materials. These blocks contain varying states of health, making it impossible to accurately analyze their composition and determine the required lithium supplementation. Without knowing the required lithium quantity, it is difficult to ensure that the regenerated cathode material will restore its performance. Additionally, after repeated use of some LIBs, phenomena such as the dissolution of TMs and the mixing of lithium with TMs occur, in addition to lithium loss. One potential solution to this issue is designing a self-saturating lithium-rich liquid-phase system that can provide an appropriate amount of lithium for re-lithiation. By predicting the total composition and elemental ratios in the mixture, sufficient transition elements can be added to the system to balance the composition. However, this approach has not been systematically researched, even at the laboratory level, and further studies may be necessary.

In summary, the recycling and repair technologies for retired NCM lithium batteries face multiple challenges, such as complex compositions, difficulty in accurately measuring lithium loss, and poor consistency in repaired materials' performance. Current mainstream technologies like solid-phase and liquid-phase repair have their advantages and drawbacks but still face issues of high energy consumption, cost, and safety in practical applications. New technologies such as electrochemical repair, though environmentally friendly, are still immature in their application to NCM materials. Moving forward, research needs to deeply explore the failure mechanisms of NCM cathode materials, develop more efficient, cost-effective, and environmentally friendly repair technologies, and establish a comprehensive system model to address key issues like battery design standardization and fault evaluation. Additionally, accurate analysis of degraded material composition and precise control of lithium supplementation must be emphasized to ensure the performance of regenerated materials. Exploring new methods like self-saturating lithium-rich liquid-phase systems could be an effective way to solve existing issues. Ultimately, efficient, economical, and safe direct repair technologies for NCM batteries will be a key competitive factor in the industry.

Direct regeneration of graphite: The future development of waste graphite repair technologies will focus on overcoming current technological bottlenecks and achieving coordinated treatment of surface and internal defects. On the one hand, low-temperature repair technologies need to extend to deep structural repairs, developing new low-temperature plasma, microwave-assisted, and other technologies to precisely repair interlayer exfoliation and microcracks while maintaining the energy-saving advantage. On the other hand, high-temperature repair technologies need to optimize energy efficiency by exploring innovative processes like instantaneous high temperature and gradient heating to shorten processing time, reduce heat loss, and combine inert gas protection to avoid material oxidation risks. Core breakthroughs include promoting multi-technology integration, such as "low-temperature surface purification + high-temperature structural repair" or "molten salt pretreatment + pulsed Joule heating," to balance surface impurity removal and lattice reconstruction; advancing intelligent control, integrating in situ characterization and AI algorithms to monitor defect evolution in real-time and dynamically optimize repair parameters for precise targeting; accelerating green process upgrades by developing nonpolluting chemical reagent systems and byproduct recycling technologies to drive low-carbon repair processes and resource recycling; and achieving continuous production by designing efficient integrated repair production lines to enhance the economic viability of large-scale applications. Through cross-scale defect coordination and energy-efficiency system optimization, waste graphite regeneration technology will continue to evolve toward low-consumption, high-efficiency, and high-value directions, providing a solid core material guarantee for the sustainable development of the lithium battery industry.

References

1 G. Ji, J. Wang, Z. Liang, K. Jia, J. Ma, Z. Zhuang, G. Zhou and H.-M. Cheng, *Nature Communications*, 2023, **14,** 584.

2 C. Cheng, X. Cao, Z. Xing and S. Tang, *Journal of Materials Science & Technology*, 2026, **241**, 262–269.

3 X. Li, Q. Zhou, X. Zhang, M. Ge, H. Zhang, Y. Yin and S.-T. Yang, *ACS Sustainable Chemistry & Engineering*, 2023, **11**, 14457–14466.

4 M. Xiao, X. Fu, M. Chen, M. Ye, C. Zhu, F. Wan and X. Guo, *ACS Applied Materials & Interfaces*, 2025, **17**, 12199–12207.

5 B. Wang, Y. Yin, H. Deng, G. Li, H. Zhu and W. He, *Chemical Engineering Journal*, 2024, **497**, 154922.

6 H. Sun, X. Li, B. Wu, K. Zhu, Y. Gao, T. Bao, H. Wu and D. Cao, *Chinese Chemical Letters*, 2025, **36,** 110041.

7 D. Guo, C. Xing, L. Chen, Z. Yang, J. Huang, G. Wang, T. Liu, M. Jiang, P. Wang and L. Fei, *Journal of Energy Chemistry*, 2025, **110**, 571–580.

8 T. Wang, C. Gao, Z. Zheng, W. Yu, M. Wang, C. Yang and J. Zhang, *Advanced Functional Materials*, 2025, DOI: 10.1002/adfm.202502930.

9 Z. Wang, H. Xu, Z. Liu, M. Jin, L. Deng, S. Li and Y. Huang, *Journal of Materials Chemistry A*, 2023, **11**, 9057–9065.

10 S. Lei, Q. Li, L. Zhang, L. Wang and Y. Yang, *Chemical Engineering Journal*, 2025, **511**, 161928.

11 Y. Wang, Y. Yang, J. Zhang, Y. Chen and C. Wang, *Applied Surface Science*, 2025, **689,** 162512.

12 J. Ma, Z. Xu, T. Yao, Z. Chen, X. Liu, Q. Sun, Y. Jin, L. Song and M. D. Zhang, *ACS Applied Materials & Interfaces*, 2024, **16**, 65119–65130.

13 J. Wang, K. Jia, J. Ma, Z. Liang, Z. Zhuang, Y. Zhao, B. Li, G. Zhou and H.-M. Cheng, *Nature Sustainability*, 2023, **6**, 797–805.

14 Y. Ji, D. Yang, Y. Li, M. Avdeev, W. H. Kan, M. Liao, Z. Zhang, W. Du, X. Pi, C. Liang, Z. Fang, X. Qi, L. Xue, R. Zhang, L. Qie and Y. Huang, *Energy Storage Materials*, 2025, **75**, 104006.

15 Z. Liu, H. Li, M. Han, L. Fang, Z. Fu, H. Zhang, G. Wang and Y. Zhang, *Advanced Energy Materials*, 2023, **13** 2302058.

16 Z. Zeng, H. Lei, X. Lu, C. Zhu, Y. Wen, J. Zhu, X. Ji, W. Sun, Y. Yang and P. Ge, *Energy Storage Materials*, 2025, **74**, 103947.

17 G. Zeng, C. Hu, R. Zhou, J. Wang, X. Liu, J. Huang, J. Yu, G. Wu, Z. Wang, C. Deng, J. He, C. Liu and X. Luo, *Energy Storage Materials*, 2025, **78**, 104287.

18 Y. Li, C. Wan, Y. Tian, J. Li, C. Yang, W. Zhang, X. Zhang, Z. Hao, Z. Yang, P. Guo, B. Yang, D. Ruan, M. Xie and J. Hu, *Applied Surface Science*, 2023, **609**, 155162.

19 Z. Zheng, D. Xie, X. Liu, H. Huang, M. Zhang and F. Cheng, *ACS Applied Materials & Interfaces*, 2024, **16**, 31137–31144.

20 X. Hu, Y.-C. Yin, C. Li, L. Zhou, L. Yang, Y. Feng, D. Zuo, C. Ning, D. Soham, S. Dai, L. Qiu, L. Zeng, H. J. Woo and J. Wan, *Science China Chemistry*, 2024, **67**, 2181–2189.

21 X. Mu, K. Huang, G. Zhu, Y. Li, C. Liu, X. Hui, M. Sui and P. Yan, *Nano Energy*, 2023, **112**, 108465.

22 S. Kumar, K. Chakravarty, K. Kumar, A. Sen, C. Ghanty, S. Chakravarty and R. Kundu, *Energy & Fuels*, 2024, **38**, 6499–6507.

23 X. Fan, C. Tan, Y. Li, Z. Chen, Y. Li, Y. Huang, Q. Pan, F. Zheng, H. Wang and Q. Li, *Journal of Hazardous Materials*, 2021, **410**, 124610.

24 J. Pu, Z. Huang, C. Wu, S. Liu, S. Liu, X. Liu, L. You, Q. Wang, Y. Zheng, R. Zeng and J. Liu, *Energy Storage Materials*, 2024, **67**, 103304.

25 M. Xiao, X. Fu, Z. Zhang, M. Ye, L. Qiu, Z. Wu, F. Wan and X. Guo, *Energy Storage Materials*, 2025, **79**, 104307.

26 S. Jin, Z. A. Lu, D. Mu, T. Lin, W. Zhang, Y. Zhang, D. Chen, R. Li and C. Dai, *Advanced Functional Materials*, 2024, **35**, 2416085.

27 P. Ge, B. Wang, Z. Zeng, C. Zhu, H. Lei, J. Li, H. Zhou, W. Sun and Y. Yang, *Advanced Functional Materials*, 2025, **35**, 2500117.

28 M. Fan, X. Chang, Y.-J. Guo, W.-P. Chen, Y.-X. Yin, X. Yang, Q. Meng, L.-J. Wan and Y.-G. Guo, *Energy & Environmental Science*, 2021, **14**, 1461–1468.

29 Z. Zhuang, J. Li, H. Ji, Z. Piao, X. Wu, G. Ji, S. Liu, J. Ma, D. Tang, N. Zheng, J. Wang and G. Zhou, *Advanced Materials*, 2024, **36**, e2313144.

30 J. Wang, J. Ma, Z. Zhuang, Z. Liang, K. Jia, G. Ji, G. Zhou and H.-M. Cheng, *Chemical Reviews*, 2024, **124**, 2839–2887.

31 T. Zhang, Z. Qin, X. Long, Y. Gao, Z. Qin, N. Zhang, C. Guan and G. Chen, *Advanced Functional Materials*, 2025, **35**, 2500608.

32 Y. Li, X. Chen, Z. Zeng, Y. Dong, S. Yuan, W. Zhao, F. Jiang, Y. Yang, W. Sun and P. Ge, *ACS Applied Energy Materials*, 2022, **5**, 7874–7888.

33 C. Xing, H. Da, P. Yang, J. Huang, M. Gan, J. Zhou, Y. Li, H. Zhang, B. Ge and L. Fei, *ACS Nano*, 2023, **17**, 3194–3203.

34 X. Liu, R. Wang, S. Liu, J. Pu, H. Xie, M. Wu, D. Liu, Y. Li and J. Liu, *Advanced Energy Materials*, 2023, **13**, 2302987.

35 B. Chen, M. Liu, S. Cao, H. Hu, G. Chen, X. Guo and X. Wang, *Journal of Alloys and Compounds*, 2022, **924**, 166487.

36 J. Yang, K. Zhou, R. Gong, Q. Meng, Y. Zhang and P. Dong, *Journal of Environmental Management*, 2023, **348**, 119384.

37 P. Xu, Q. Dai, H. Gao, H. Liu, M. Zhang, M. Li, Y. Chen, K. An, Y. S. Meng, P. Liu, Y. Li, J. S. Spangenberger, L. Gaines, J. Lu and Z. Chen, *Joule*, 2020, **4**, 2609–2626.

38 K. Jia, J. Wang, Z. Zhuang, Z. Piao, M. Zhang, Z. Liang, G. Ji, J. Ma, H. Ji, W. Yao, G. Zhou and H.-M. Cheng, *Journal of the American Chemical Society*, 2023, **145**, 7288–7300.

39 P. Xu, D. H. S. Tan, B. Jiao, H. Gao, X. Yu and Z. Chen, *Advanced Functional Materials*, 2023, **33**, 2213168.

40 Y. Cao, J. Li, H. Ji, X. Wei, G. Zhou and H.-M. Cheng, *Energy Storage Materials*, 2024, **70**.

41 D. Tang, G. Ji, J. Wang, Z. Liang, W. Chen, H. Ji, J. Ma, S. Liu, Z. Zhuang and G. Zhou, *Advanced Materials*, 2023, **36**, 2309722.

42 K. Jia, Y. He, Z. Cao, Z. Piao, M. Zhang, J. Gao, J. Wang, A. M. Abdelkader, Z. Liang, R. V. Kumar, G. Yang, S. Ding, G. Zhou, Y.-L. He and K. Xi, *Science China Chemistry*, 2025, DOI: 10.1007/s11426-025-2565-5.

43 J. Luo, J. Zhang, Z. Guo, Z. Liu, C. Wang, H. Jiang, J. Zhang, L. Fan, H. Zhu, Y. Xu, R. Liu, J. Ding, Y. Chen and W. Hu, *Advanced Materials*, 2024, **36**, 2405956.

44 K. Jia, J. Ma, J. Wang, Z. Liang, G. Ji, Z. Piao, R. Gao, Y. Zhu, Z. Zhuang, G. Zhou and H. M. Cheng, *Advanced Materials*, 2022, **35**, 2208034.

45 Y. Liu, J. Bai, R. Shi, P. Wang, K. Xiao, S. Wang, S. Qiu, X. Wang, X. Zhu, K. Yang, G. Zhou, B. Zhao and Y. Sun, *Advanced Materials*, 2025, DOI: 10.1002/adma.202511246.

46 Y. Lan, X. Li, G. Zhou, W. Yao, H. M. Cheng and Y. Tang, *Advanced Science*, 2023, **11**, 2304425.

47 L. Zhang, H. Gao, Y. Zhu, I. Tran, W. Tang, J. Lin, A. U. Mu, J. Wu, W. Li, D. Nordlund, L. Mu and Z. Chen, *Advanced Energy Materials*, 2025, DOI: 10.1002/aenm.202406084.

48 Z. Fei, Y. Xing, P. Dong, Q. Meng and Y. Zhang, *JOM*, 2023, **75**, 3632–3642.

49 J. Wang, Q. Zhang, J. Sheng, Z. Liang, J. Ma, Y. Chen, G. Zhou and H.-M. Cheng, *National Science Review*, 2022, **9**,.097.

50 Y. Wang, H. Yu, Y. Liu, Y. Wang, Z. Chen, D. Tang, W. Li and J. Li, *Electrochimica Acta*, 2022, **407**, 139863.

51 N. Zhang, W. Deng, Z. Xu and X. Wang, *Carbon Energy*, 2022, **5**, 231.

52 Z. Zhang, W. He, G. Li, J. Xia, H. Hu, J. Huang and S. Zhang, *ECS Electrochemistry Letters*, 2014, **3**, A58–A61.

53 Y. Shi, G. Chen and Z. Chen, *Green Chemistry*, 2018, **20**, 851–862.

54 S. Zhao, W. Zhang, G. Li, H. Zhu, J. Huang and W. He, *Resources, Conservation and Recycling*, 2020, **162**, 1050019.

55 Y. Liu, H. Yu, Y. Wang, D. Tang, W. Qiu, W. Li and J. Li, *Waste Management*, 2022, **143**, 186–194.

56 P. Xu, Z. Yang, X. Yu, J. Holoubek, H. Gao, M. Li, G. Cai, I. Bloom, H. Liu, Y. Chen, K. An, K. Z. Pupek, P. Liu and Z. Chen, *ACS Sustainable Chemistry & Engineering*, 2021, **9**, 4543–4553.

57 V. Gupta, X. Yu, H. Gao, C. Brooks, W. Li and Z. Chen, *Advanced Energy Materials*, 2022, **13**, 2203093.

58 X. Yu, S. Yu, Z. Yang, H. Gao, P. Xu, G. Cai, S. Rose, C. Brooks, P. Liu and Z. Chen, *Energy Storage Materials*, 2022, **51**, 54–62.

59 T. Wang, H. Luo, Y. Bai, J. Li, I. Belharouak and S. Dai, *Advanced Energy Materials*, 2020, **10**, 2001204.

60 B. K. Biswal, B. Zhang, P. Thi Minh Tran, J. Zhang and R. Balasubramanian, *Chemical Society Reviews*, 2024, **53**, 5552–5592.

61 Z. Zeng, H. Lei, J. Li, B. Wang, S. Lei, X. Ji, W. Sun, Y. Yang and P. Ge, *Chemical Engineering Journal*, 2024, **499**, 155616.

62 Z. Zeng, P. Xu, J. Li, C. Yi, W. Zhao, W. Sun, X. Ji, Y. Yang and P. Ge, *Advanced Functional Materials*, 2023, **34**, 2308671.

63 K. Jia, G. Yang, Y. He, Z. Cao, J. Gao, H. Zhao, Z. Piao, J. Wang, A. M. Abdelkader, Z. Liang, R. V. Kumar, G. Zhou, S. Ding and K. Xi, *Advanced Materials*, 2024, **36**, 2313273.

64 M. Du, J.-Z. Guo, S.-H. Zheng, Y. Liu, J.-L. Yang, K.-Y. Zhang, Z.-Y. Gu, X.-T. Wang and X.-L. Wu, *Chinese Chemical Letters*, 2023, **34**, 107706.

65 S. Lei, J. Li, L. Hu, D. Qiao, W. Sun, P. Ge and Y. Yang, *Journal of Energy Chemistry*, 2024, **99**, 458–465.

66 X. Yue, Y.-X. Yao, J. Zhang, S.-Y. Yang, W. Hao, Z. Li, C. Tang, Y. Chen, C. Yan and Q. Zhang, *Angewandte Chemie International Edition*, 2025, **64**, e202413926.

67 W. Zhong, R. He, L. Peng, W. Liu, J. Peng, H. Zhu, J. Xiang, S. Cheng and J. Xie, *Advanced Energy Materials*, 2025, **15**, 2406007.

68 M. Fan, Q. Meng, X. Chang, C.-F. Gu, X.-H. Meng, Y.-X. Yin, H. Li, L.-J. Wan and Y.-G. Guo, *Advanced Energy Materials*, 2022, **12**, 2103630.

69 R. Malik, F. Zhou and G. Ceder, *Nature Materials*, 2011, **10**, 587–590.

70 H. Moon, D. Kim, G. Park, K. Shin, Y. Cho, C. Gong, Y.-S. Lee, H. Nam, S. Hong and N.-S. Choi, *Advanced Functional Materials*, 2024, **34**, 2403261.

71 H. Du, Y. Wang, Y. Kang, Y. Zhao, Y. Tian, X. Wang, Y. Tan, Z. Liang, J. Wozny, T. Li, D. Ren, L. Wang, X. He, P. Xiao, E. Mao, N. Tavajohi, F. Kang and B. Li, *Advanced Materials*, 2024, **36**, 2401482.

72 J. C. Hestenes and L. E. Marbella, *ACS Energy Letters*, 2023, **8**, 4572–4596.

73 Z. Qin, T. Zhang, X. Gao, W. Luo, J. Han, B. Lu, J. Zhou and G. Chen, *Advanced Materials*, 2024, **36**, 2307091.

74 C.-F. Gu, X. Chang, S. Xiao, Z.-Y. Zhou, C. Li, B. Yuan, Q. Meng and Y.-G. Guo, *Advanced Materials*, 2025, **37**, e10189.

75 M. Tian, Y. Yan, H. Yu, L. Ben, Z. Song, Z. Jin, G. Cen, J. Zhu, M. Armand, H. Zhang, Z. Zhou and X. Huang, *Advanced Materials*, 2024, **36**, 2400707.

76 Y. Song, H. Zhang, S. Guo, C. Lin, Z. Wang, X. Hu, M. Cao, L. Qie, D. Li and X. Ji, *Nano Research*, 2025, **18**, 94907598.

77 W. Zhao, S. Lei, J. Li, F. Jiang, T. Wu, Y. Yang, W. Sun, X. Ji and P. Ge, *Advanced Energy Materials*, 2024, **14**, 2304431.

78 D. Kang, J. Im, S. Chae, H. Jang and M. Ko, *Energy & Environmental Materials*, 2025, **9**, e70037.

79 J. Wang, Z. Qi, Q. Zhao and K. Shih, *Advanced Science*, 2025, **12**, e04609.

80 N. Zhang, H. Li, C. Ye and S.-Z. Qiao, *Advanced Materials*, 2025, **37**, 2504380.

81 S. Zhang, Y. Liu, G. Chen, Z. M. El-Bahy, D. A. Alshammari, M. H. Helal, J. Zhou and J. Han, *Advanced Materials*, 2025, **38**, e13547.

82 J. G. Sederholm, A. Patra, Z. Liu, J.-W. Lin, C. Juarez-Yescas, P. Wang and P. V. Braun, *Advanced Functional Materials*, 2026, **36**, e11009.

83 A. De la Hoz, A. Diaz-Ortiz and A. Moreno, *Chemical Society Reviews*, 2005, **34**, 164–178.

84 Y. Liu, X. Zhang, W. Ma, H. Duan, Q. Zhao and Z. Liang, *Journal of Energy Storage*, 2024, **99**, 113407.

85 Z. Wang, Y. Chen, F. Zhou, R. Qin, Y. Tian, Z. Xue and T. Mu, *Green Chemistry*, 2024, **26**, 5988–5996.

86 T. Xu, X. Li, Y. Wang and Z. Tang, *ACS Applied Materials & Interfaces*, 2023, **15**, 14184–14195.

87 L. Chen, C. Xing, Z. Yang, S. Tao, Y. Zhang, G. Wang, P. Yang, J. Song, J. Chen and L. Fei, *Advanced Functional Materials*, 2025, **35**, 2411182.

88 W. Wang, R. Zhan, Y. Li, Z. Chen, R. Feng, Y. Tan, X. Duan, J. Wang, Y. Lu and Y. Sun, *ACS Energy Letters*, 2025, **10**, 1577–1584.

89 Y. Yang, S. L. Song, S. Y. Lei, W. Sun, H. S. Hou, F. Jiang, X. B. Ji, W. Q. Zhao and Y. H. Hu, *Waste Management*, 2019, **85**, 529–537.

90 H. R. Da, M. Gan, D. F. Jiang, C. X. Xing, Z. Y. Zhang, L. F. Fei, Y. J. Cai, H. T. Zhang and S. J. Zhang, *ACS Sustainable Chemistry & Engineering*, 2021, **9**, 16192–16202.

91 S. Y. Xie, Y. Dong, X. Wang, Z. H. Zeng, H. Y. Zhou, Z. Q. Yuan, W. Sun, X. B. Ji, Y. Yang and P. Ge, *Energy Storage Materials*, 2024, **70**, 15.

92 X. Li, C. Deng, M. Liu, J. Xiong, X. Zhang, Q. Yan, J. Lin, C. Chen, F. Wu, Y. Zhao, R. Chen and L. Li, *eScience*, 2025, **5**, 100394.

93 C. X. Yi, P. Ge, X. Q. Wu, W. Sun and Y. Yang, *Journal of Energy Chemistry*, 2022, **72**, 97–107.

94 H. Da, S. S. Pan, J. Li, J. Q. Huang, X. D. Yuan, H. T. Dong, J. X. Liu and H. T. Zhang, *Energy Storage Materials*, 2023, **56**, 457–467.

95 H. Shang, X. Hao, Y. Zhou, J. Peng, L. Guo, H. Li and B. Sun, *Nanomaterials*, 2025, **15**, 658.

96 T. Y. Li, L. Tao, L. Xu, T. T. Meng, B. C. Clifford, S. K. Li, X. P. Zhao, J. C. Rao, F. Lin and L. B. Hu, *Advanced Functional Materials*, 2023, **33**, 8.

97 Z. Shang, Z. Naizhe, Z. W. Ying, D. J. Zou, F. Dai, J. H. Xie, J. Shao, X. G. Liu and S. M. Xu, *Chemical Engineering Journal*, 2025, **505**, 10.

98 S. Yuan, J. Li, W. Xia, Z. Li, W. Tian, Y. Yang, X. Ji and P. Ge, *Advanced Functional Materials*, 2024, **34**, 2409295.

99 Y. S. Ji, H. Zhang, D. Yang, Y. J. Pan, Z. L. Zhu, X. Q. Qi, X. P. Pi, W. C. Du, Z. H. Cheng, Y. G. Yao, L. Qie and Y. H. Huang, *Advanced Materials*, 2024, **36**, 11.

6

Pollution Prevention and Life Cycle Assessment of Degraded Lithium-ion Battery Recycling

Congrui Ouyang, Jin Wang, Wenhao Yu

6.1 Pollution Prevention

Lithium-ion batteries (LIBs) are the core components of new energy vehicles (NEVs) and energy storage systems (ESSs). Their widespread applications have resulted in a sharp increase in degraded batteries. Improper treatment of these degraded batteries can release hazardous substances, such as heavy metals and decomposition products of organic electrolytes, which may cause serious environmental risks. Therefore, reducing such risks during large-scale treatment of degraded batteries has become a challenge.[1, 2]

At present, three main recycling routes are employed, which are pyrometallurgical recycling, hydrometallurgical recycling, and direct regeneration recycling (Figure 6.1).[3] Although these methods have achieved moderate pollution prevention, considerable gaps remain, particularly regarding recycling efficiency and overall environmental performance. With the increasing demand for cleaner recycling, policy and technological efforts are being directed worldwide toward greener solutions, underscoring the urgent need to develop more sustainable recycling processes.

This chapter focuses on pollution prevention of typical degraded battery materials recycling, including lithium iron phosphate (LFP), lithium nickel cobalt manganese oxide (NCM), lithium nickel cobalt aluminum oxide (NCA), and lithium cobalt oxide (LCO). For each recycling route, potential sources of pollution are identified, and corresponding prevention measures are discussed. This systematic comparison not only reveals the limitations of existing technologies but also clarifies the areas where improvements are most needed. Finally, possible future directions are outlined to provide guidance for cleaner and more efficient recycling strategies.

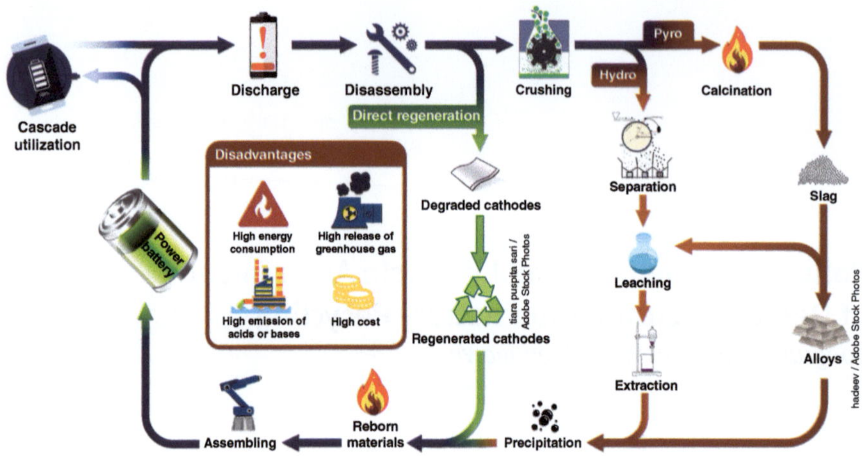

Figure 6.1 Different recycling routes for degraded LIBs.

6.1.1 Current Status of Pollution and Prevention Strategies in Degraded Battery Recycling

The main components of LIBs, including active materials (such as LFP, NCM, NCA, and LCO), electrolytes (with variations in lithium salts, solvents, and additives), and binders (such as F-containing organics), exhibit significant differences in chemical composition. Consequently, the types, forms, and environmental risks of pollutants generated during the recycling of different batteries vary considerably. The diversity in composition emphasizes the importance of implementing customized pollution prevention strategies for different materials.

In response, this section provides a systematic examination of pollution prevention in mainstream recycling processes, including pyrometallurgical recycling, hydrometallurgical recycling, and direct regeneration recycling. The discussion is guided by existing policies, regulatory frameworks, technical standards, and regulations. For each route, typical pollutants are identified, and the mechanisms responsible for specific contamination are analyzed in detail. Examples include heavy metal-rich dust and toxic gases such as hydrogen fluoride in pyrometallurgical processes; concentrated acidic effluents, heavy metal leachates, and residual organic extractants in hydrometallurgical processes; and toxic fluorinated gases along with metallic impurities such as aluminum and copper in direct regeneration processes. Major pollution prevention strategies are then summarized, including advanced gas purification systems, wastewater treatment technologies, and resource recovery technologies aimed at minimizing pollutant generation at the source.

To address growing environmental concerns, governments worldwide have recently implemented a range of regulations and standards to enhance pollution prevention for battery recycling,[4] as illustrated in Figure 6.2. In China, a mandatory regulatory framework has been established across the entire battery recycling chain. The "Emission Standards of Pollutants for Battery Industry" were issued in 2013, followed by the "Industry Standards for Comprehensive Utilization of Degraded

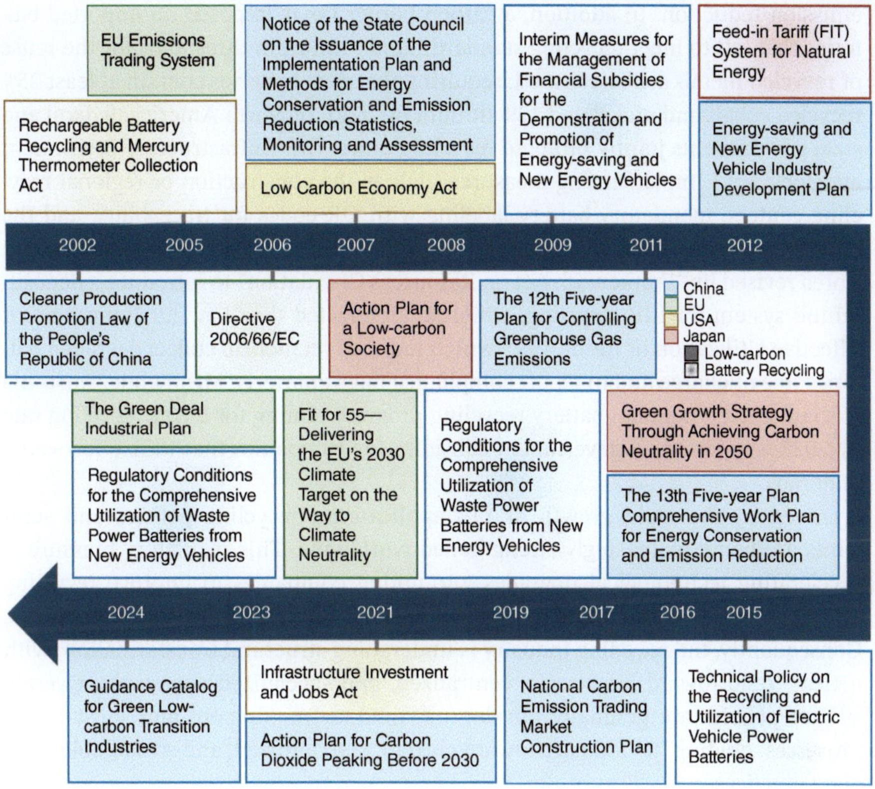

Figure 6.2 Evolution of global policies on the low-carbon development of the battery recycling industry.

NEV Power Batteries" in 2019 and its revised version in 2024. These standards stipulate comprehensive recovery rates of over 90% for lithium and over 98% for nickel, cobalt, and manganese. They also establish limits on the discharge of wastewater, waste gas, noise, and solid waste, specifying thresholds for main pollutants, including fluorides and heavy metals. Recycling companies are required to meet these standards through stricter environmental management. In addition, emission permit systems have been implemented in some regions, requiring companies to install online monitoring equipment, such as continuous emission monitoring systems (CEMS) for flue gas and automatic water quality analyzers, with data transmitted in real time to provincial ecological platforms. An extended producer responsibility (EPR) has also been introduced, which includes a recycling fund levied according to battery capacity, certification of benchmark companies, a "whitelist" policy, and the elimination of outdated capacity exceeding 400,000 tons per year.

Globally, the EU has established a life-cycle-oriented green regulatory framework through a series of legislative measures, including the EU Battery Regulation 2023/1542 and the Industrial and Livestock Rearing Emissions Directive (IED 2.0), which set clear requirements for recycling efficiency, mandatory disclosure of carbon footprint data, and the application of best available techniques for emission reduction. In addition, a carbon border tax is imposed on imported batteries that fail to meet recycling standards, and minimum thresholds for the reuse of recycled metals are established, requiring that new batteries contain at least 25% recycled cobalt and nickel and 12% lithium by 2030. In North America, federal and state governments jointly promote recycling under the "Infrastructure Investment and Jobs Act," implementing measures such as the construction of regional recycling centers, mandatory battery labeling with QR codes for traceability, and the establishment of statewide monitoring systems. In the Asia-Pacific region, South Korea revised its "Framework Act on Resources Circulation" to introduce a deposit-refund system for LIB recycling, while Japan enacted the "Law for Promotion of Effective Utilization of Resources," which requires retailers to collect degraded batteries free of charge and to report compliance rates quarterly. Meanwhile, Australia has launched a national battery recycling program aiming for a 60% recycling rate by 2025, supported by government subsidies for companies that utilize recovered materials.

Driven by the rapid growth of LIB applications, recycling policies and standards are being increasingly strengthened worldwide. This regulatory pressure is accelerating technological upgrades, prompting companies to improve recycling efficiency and to develop and implement advanced pollution prevention methods. Consequently, the recycling industry is undergoing structural transformation, with a clear trend toward large-scale, centralized, and specialized operations. Overall, stricter policies are gradually phasing out outdated equipment and substandard processes, guiding the industry toward cleaner, standardized, and sustainable recycling practices.

6.1.2 Pollutants and Pollution Prevention Strategies of Different Recycling Processes

6.1.2.1 Pyrometallurgical Recycling

Pyrometallurgical recycling extracts valuable metals from degraded LIBs through high-temperature treatment, during which the metals are melted, volatilized, or alloyed. After deep discharge and dismantling, the batteries are crushed and subjected to high-temperature smelting at 700–1600 °C, with the active materials mixed with additives such as limestone and coke. Flammable components, including conductive carbon, polyvinylidene fluoride (PVDF) binders, and electrolytes, are burned, releasing energy that sustains the smelting process. High-value metals such as cobalt, nickel, and copper are incorporated into the metallic alloy, while most lithium, aluminum, and manganese are either retained in the slag or volatilized. After cooling, the metallic alloy and slag separate into different layers and are collected as products, with the slag able to undergo further chemical extraction to recover residual lithium (Figure 6.3).[5, 6] The pyrometallurgical process is associated with significant pollutant emissions, including hazardous gases, dust, heavy-metal-containing slag, and liquid wastes. This section summarizes the types of pollutants generated during pyrometallurgical recycling and the current state of prevention strategies.

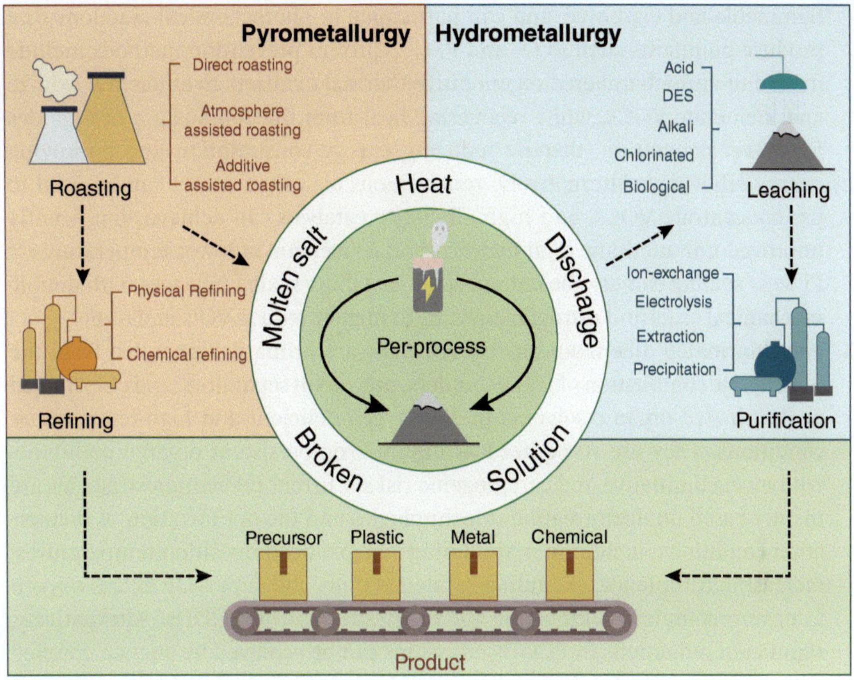

Figure 6.3 Schematic of the main processes in pyrometallurgical and hydrometallurgical recycling of degraded LIBs. *Source:* [7] / John Wiley & Sons/ CC BY 4.0.

1. F-containing gases, such as HF and PF_5, are mainly derived from the decomposition of battery electrolyte components (e.g., $LiPF_6$ and other F-containing lithium salts) and PVDF binders at high temperatures. HF is highly corrosive, posing significant hazards to both operators and equipment. PF_5 further reacts with water to generate additional HF and release heat, resulting in severe secondary pollution.[8] Current pollution prevention strategies are predominantly based on multistage alkaline spray tower, in which alkaline slurries such as $Ca(OH)_2$ are employed as the primary absorbent, and HF is captured through the formation of insoluble CaF_2.[9] Additionally, Na_2CO_3 solution can be added in the final stage for deep purification, further reducing fluoride ion concentrations to meet the relevant standard requirements. Alternatively, nanoscale Al_2O_3 powder with an ultrahigh specific surface area can be introduced, allowing HF to be removed through dry adsorption followed by the formation of AlF_3.[10] The potential hazard of Al_2O_3 dust is controlled using pulse jet bag filters. In addition, dielectric barrier discharge (DBD) using low-temperature plasma technology can be applied, allowing F-containing gases to be dissociated into radicals at relatively low temperatures, followed by synergistic treatment in denitrification systems to achieve their decomposition and removal.[11]
2. Volatile organic compounds (VOCs) are primarily generated from the thermal decomposition of organic carbonate solvents in electrolytes (e.g., EC, DMC, and DEC). VOCs, including methane, ethylene, and benzene derivatives, are

flammable and explosive, and can participate in photochemical reactions that produce pollutants such as O_3 and $PM_{2.5}$. Current prevention methods include installing multichambered regenerative thermal oxidizers to efficiently oxidize and incinerate VOCs, while recovering heat from the combustion of organics for power generation, thereby reducing energy consumption and improving energy efficiency. Alternatively, zeolite molecular sieve rotors can be used to pre-concentrate VOCs, and high-efficiency catalysts can achieve significantly improved non-methane total hydrocarbon conversion at lower temperatures.[12] Process sealing can also be enhanced by installing sealed furnaces with double mechanical seals and nitrogen curtains to further reduce VOC emissions.

3. Polychlorinated dibenzo-p-dioxins (PCDDs) are primarily generated from the incomplete combustion of PVDF binders, polyolefin separators, and chlorinated or fluorinated organic solvents under oxygen-deficient and high-temperature conditions. They are recognized as highly toxic persistent organic pollutants with bioaccumulative and carcinogenic risks. Current prevention strategies are mainly based on deep-treatment technologies and the optimization of incineration conditions, including maintaining suitable decomposition temperatures, increasing turbulence, extending residence time, and supplying excess oxygen to ensure complete combustion for the destruction of PCDDs. Alternatively, significant reductions in PCDDs emissions can be achieved by quench coupled with adsorption technology, where atomized water is sprayed to rapidly cool the flue gases, thereby preventing the resynthesis of PCDDs, while activated carbon injection combined with high-efficiency bag filters ensures their effective adsorption and removal.
4. Carbon oxides and nitrogen oxides are major gaseous emissions, among which CO_2 and CO are mainly produced from the combustion of carbon-based materials (e.g., conductive additives, PVDF, and coke), while NO_x is primarily generated from the oxidation of atmospheric nitrogen at high temperatures. CO_2 is a critical greenhouse gas, CO exhibits strong affinity for hemoglobin in blood and poses severe poisoning risks, and NO_x can react with VOCs to form O_3 that damages the respiratory system. Current prevention strategies include high-efficiency carbon capture using chemical chain combustion and low-cost, high-efficiency catalysts such as Fe_2O_3 to convert CO_2 into high-value chemicals;[13] CO can be treated via oxy-fuel combustion coupled with heat recovery for power generation. For NO_x prevention strategies, selective catalytic reduction (SCR) and selective non-catalytic reduction (SNCR) coupled technologies are widely employed, with additional flue gas denitrification units, using high-temperature urea solution injection and catalysts for efficient NO_x conversion.
5. Heavy metal-containing slags are mainly produced from pyrometallurgical recycling, in which Li, Mn, Al, and parts of Fe and Cu exist in oxide or complex salt forms.[14] When exposed to rainwater, these slags can undergo leaching, resulting in the migration of heavy metals into soil and groundwater. Current prevention strategies include stabilizing the slags with cement to inhibit leaching, or employing low-cost fluxes (SiO_2-Al_2O_3-CaO) at high temperatures to

form non-flowing glassy phases and thereby prevent heavy metal migration, which can be coupled with reduction roasting to recover high-value metals.[15] Alternatively, toxic metals can be leached with acids to reduce slag hazards, or slags can be fractionated to enhance resource recovery. Li can be recovered via sulfate maturation followed by water leaching, Al through alkali fusion and acid leaching, Cu by ammonia leaching and electrodeposition, and Mn and Fe through acid leaching and precipitation.

6. Dust is primarily generated during crushing, feeding, smelting, and slag discharge processes, and often contains metallic Li, Ni, Co, Cu, or their oxides. Inhaled dust can deposit in human lungs, causing chronic diseases. Heavy metal dust is also persistent in the environment and can exert long-term ecological impacts. Current prevention strategies include the installation of negative-pressure dust removal systems throughout pyrometallurgical operations, such as negative-pressure crushing and feeding, cyclone pre-dust removal, bag filters combined with electrostatic precipitation, high-temperature coated filter bags, and wet electrostatic dust removal, all of which aim to minimize dust emissions. In addition, automated sealed systems are employed in operational areas to further prevent dust dispersion.
7. Washing wastewater is mainly generated from gas purification and slag leaching processes. It contains high concentrations of salts such as NaF, Na_2SO_4, and metal chlorides. Discharge of this wastewater can cause water eutrophication and toxic accumulation. Current prevention strategies involve using neutralization pools to convert residual acids and bases into salts, followed by multistage sedimentation and filtration to remove high concentrations of salts. Membrane separation technologies, such as nanofiltration, are applied to selectively capture ions, while reverse osmosis is employed to concentrate and recover valuable components, including unreacted acids/bases and fluorine-containing products. Useful compounds such as NaF and Na_2SO_4 are crystallized by forced-circulation evaporators. Moreover, bipolar membrane electrodialysis is utilized for the recovery and regeneration of acids and bases.
8. Cooling water and leachate mainly originate from furnace cooling water system leaks or the reaction of rainwater with metal powders in raw material piles. Leachates contain high concentrations of soluble metal ions, potentially causing long-term soil and water pollution. Current prevention strategies include engineering liners with multilayer systems such as high-density polyethylene impermeable (HDPE) membranes and bentonite mats to prevent leakage. Cooling water is controlled by a closed-loop system with regular blowdown treatment. A rainwater-wastewater diversion system and wastewater collection ponds are established to achieve a dual barrier of seepage prevention and treatment, ensuring that wastewater is not discharged externally (Figure 6.4).

6.1.2.2 Hydrometallurgical recycling

Hydrometallurgical recycling has become one of the mainstream technologies for degraded batteries due to its high selectivity, ability to produce high-purity products, and suitability for recovering metals at low concentrations. It primarily relies on

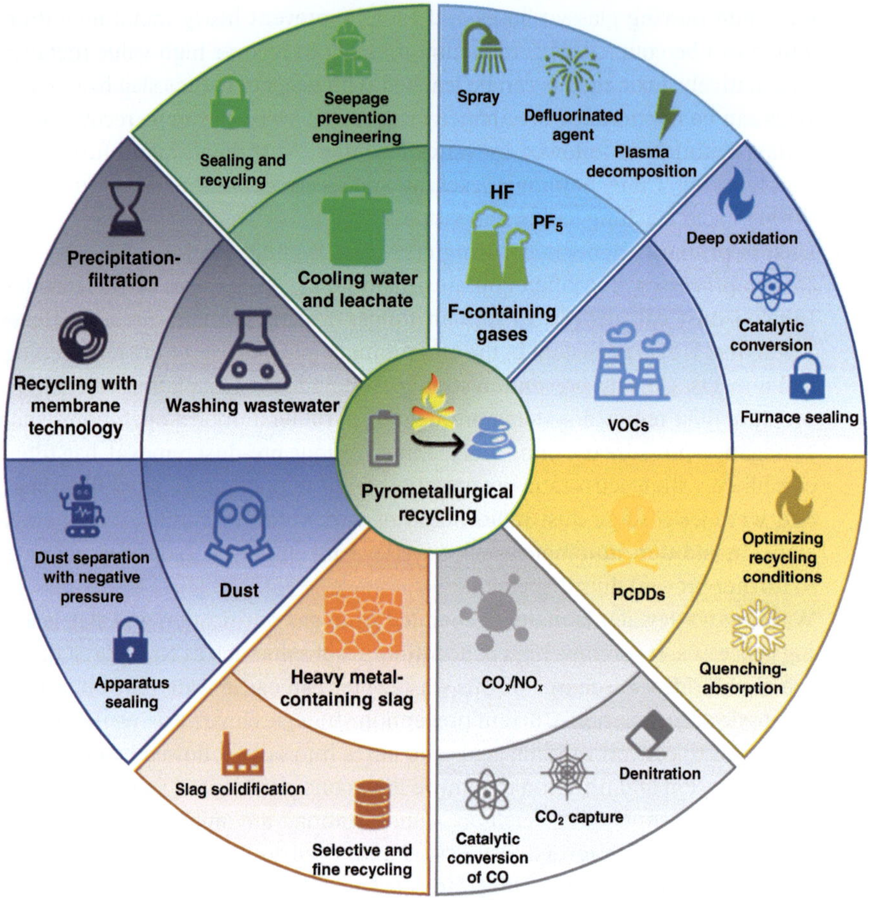

Figure 6.4 Summary of the pollutants and prevention strategies in pyrometallurgical recycling.

water-based chemical processes such as acid leaching, solvent extraction, precipitation, and electrolysis to extract lithium, cobalt, nickel, and manganese from cathode materials.[16] In this process, batteries are first discharged, dismantled, crushed, and sieved, followed by acid leaching. Impurities are removed through precipitation, valuable metals are separated using solvent extraction, and finally recovered via precipitation. Wastewater and residues are then treated to ensure safe recycling (Figure 6.3). This process involves extensive chemical reactions and the use of liquid media, leading to the generation of significant pollutants, such as acid residual, wastewater with high-concentration salts, and organic extract regent's residual.[17] The following section summarizes the main pollutants and their prevention strategies in hydrometallurgical recycling.

1. **Acid residual:** Strong acids, such as H_2SO_4 and HCl, are commonly used for metal leaching. Residual acids remaining after leaching are highly acidic, which can pollute water bodies and reduce their buffering capacity. Strong acidity can

also mobilize heavy metals deposited in ecosystems, causing significant environmental impacts. Current prevention strategies include multistage neutralization and precipitation systems using alkaline reagents such as lime or NaOH to neutralize acid residual. Further evaporation, concentration, or membrane treatment is applied to remove salts and recover valuable by-products.[18]

2. **Wastewater with high-concentration salts:** These wastewaters, containing elevated concentrations of Na, Ca, Li, SO_4^{2-}, Cl^-, and NO_3^-, are commonly generated during neutralization, extraction, and washing processes. Such waste is difficult to treat biologically, and its continuous discharge can lead to soil and water salinization. The high conductivity of the solution also increases the risk of equipment corrosion. Prevention strategies are applied by combining membrane concentration and evaporation-crystallization to recover pollutants.[19] Lithium can also be selectively recovered from wastewater, and water reused within closed loops.
3. **Organic extract regent's residual:** Organic extractants such as P507, TBP, and N235 are usually used in metal separation. Residual extractants are volatile and flammable, and may lead to the formation of O_3 and fine particulate matter through photochemical reactions. They are poorly biodegradable, and heavy metals can be adsorbed onto these residues, thereby complicating subsequent treatment. Prevention strategies are implemented through oil-water separation, followed by activated carbon adsorption to recover extractants, as well as solvent regeneration via back-extraction and membrane separation. Water-soluble extractants, such as hydrogen peroxide- or citric acid-based systems, are also employed to eliminate organic pollution.[20]
4. **Slag after leaching:** Undissolved materials such as aluminum foil, carbon powder, and separator fragments remain in leaching residues after leaching. Neutralization of the acidic leachate can result in the formation of insoluble metal fluorides, such as CaF_2, further increasing the volume of residues. Fluoride contained in the residues may be leached into water by rainfall, thereby posing environmental risks. Prevention strategies are implemented through classification using toxicity characteristic leaching procedures, as well as the recovery and reuse of recyclable materials, such as carbon powder and aluminum foil. Nonrecyclable residues are subjected to thermal defluorination or chemical conversion, followed by further purification for subsequent use in metallurgy, optical glass manufacturing, or the production of high-value fluorinated compounds.
5. Sludges rich in nickel hydroxide, cobalt hydroxide, and manganese dioxide are generated during neutralization and precipitation. These sludges are concentrated in heavy metals, prone to cracking and dusting, and may contain residual extractants, thereby presenting potential hazards. Prevention strategies are applied through filter pressing and low-temperature drying to form stable cakes, which are then safely disposed of or reused as metallurgical feedstock. Sludges containing high concentrations of nickel or cobalt can be further processed for

metal recovery, whereas sludges contaminated with organic compounds should be incinerated to prevent the release of pollutants.

6. **Volatile gases:** VOCs are generated during the hydrometallurgical recovery process due to the volatilization of extractants, mixing and stirring of liquids, and leakage from organic liquid storage. These compounds, including benzene, ethers, and phosphate esters, exhibit strong toxicity, and long-term inhalation can affect the nervous system as well as liver and kidney function, posing risks to human health. Prevention strategies are implemented through the use of sealed reactors combined with negative-pressure ventilation systems to prevent the escape of volatile gases and ensure timely removal of pollutants. Escaped gases can also be treated and recovered by installing activated carbon or condensation recovery towers. Additionally, conventional organic extractants can be substituted with low-volatility or water-soluble extractants to fundamentally eliminate hazards.
7. **Corrosive gases:** During the leaching process, the used HCl and HNO_3 may lead to the generation of corrosive gases. These gases can severely corrode metal pipelines and equipment and pose direct health risks to operators. Prevention strategies are implemented through the installation of neutralization absorption towers to reduce gas emissions, along with the adjustment and optimization of reagents and formulations to improve reaction conditions and minimize the use of volatile solvents, thereby suppressing the generation of corrosive gases at the source. Furthermore, online gas monitoring and alarm systems are deployed to enable real-time detection of gas escape and ensure the timely treatment and safe discharge of corrosive gases.
8. **Cooling water and leachate:** Treatment methods are similar to those in pyrometallurgical recycling, including engineered liners, closed-loop cooling, rainwater and wastewater separation, and collection for safe treatment (Figure 6.5).

6.1.2.3 Direct Regeneration Recycling

Currently, direct regeneration recycling remains at the laboratory scale but has increasingly attracted attention in the field of battery recycling. Unlike pyrometallurgical or hydrometallurgical processes, direct regeneration specifically addresses the failure mechanisms of degraded battery materials without decomposing them into metal alloys or ions in solution. It restores crystal structures and repairs materials through methods such as high-temperature lithium supplementation sintering or liquid-phase lithium supplementation followed by sintering. By avoiding high energy consumption and the use of strong acids or bases, this approach provides a more energy-efficient and environmentally friendly alternative.[21, 22] Nevertheless, this method still leads to the generation of certain pollutants during the recycling process.

1. **Wastewater after electrode peeling:** To minimize impurities and preserve the performance of regenerated materials, laboratory procedures are typically conducted using liquid reagents to strip active materials from electrodes. For example, deionized water can be used to strip graphite anodes and LFP cathodes, while low-toxicity organic solvents such as deep eutectic solvents[23] or phytic acid[24, 25] are used for NCM, NCA, or LCO cathodes. The stripping

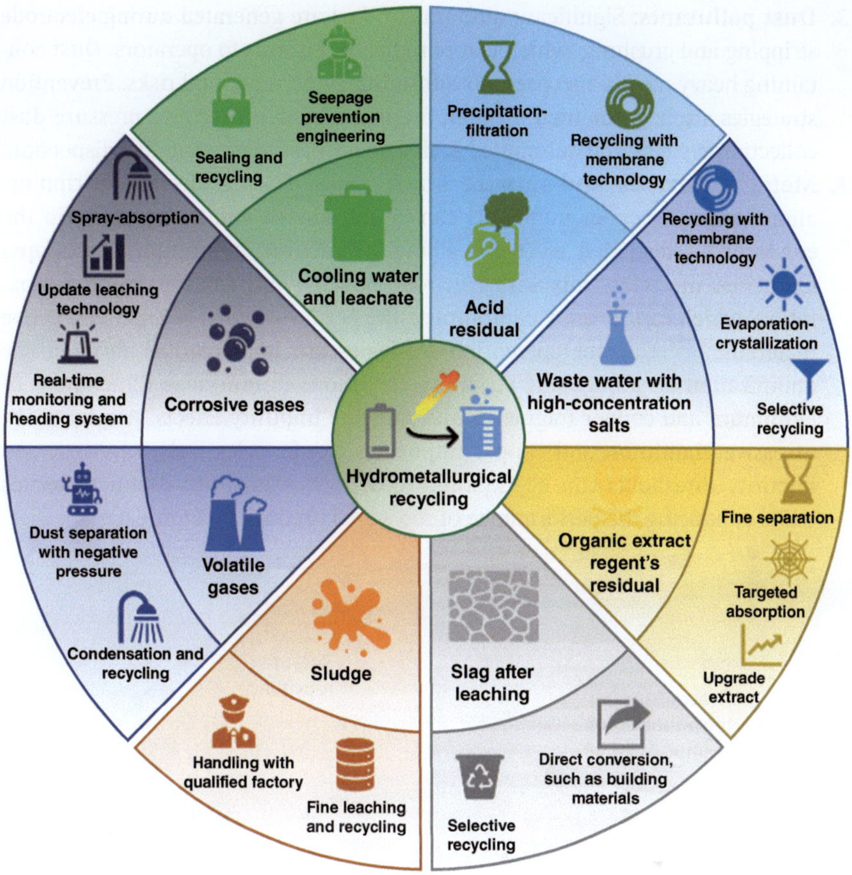

Figure 6.5 Summary of the pollutants and prevention strategies in hydrometallurgical recycling.

process results in the generation of waste liquids enriched with phosphorus, fluorine, heavy-metal ions, and toxic organic reagents. Prevention strategies are implemented through the selective recovery of valuable components, such as lithium, from the waste liquid, pre-treatment of residual electrolytes to prevent water contamination, or the use of biocompatible, low-toxicity aqueous solvents to reduce the generation of highly toxic waste.

2. **F-containing gases, VOCs, and PCDDs:** These pollutants are primarily generated from the decomposition of F-containing lithium salts, such as $LiPF_6$, carbonate-based electrolyte solvents, and PVDF binders during electrode stripping or high-temperature treatment. Current prevention strategies are applied by optimizing the electrode stripping process to reduce electrolyte exposure and recover valuable components for reuse, thereby minimizing pollution while improving economic efficiency. Alternatively, high-temperature decomposition conditions can be optimized to limit the generation of gaseous pollutants, while toxic gases are efficiently captured and rapidly decomposed to prevent their release.[26]

3. **Dust pollutants:** Significant amounts of dust are generated during electrode stripping and crushing, which can pose health hazards to operators. Dust containing heavy metals also presents substantial environmental risks. Prevention strategies are implemented through the installation of negative-pressure dust collection systems and automated sealed enclosures to prevent dust dispersion.
4. **Metal impurities and impure black mass:** During electrode stripping, aluminum and copper fragments can be introduced due to variations in the condition of degraded electrode storage. Moreover, aluminum and copper impurities inevitably mix with electrode powders during crushing and separation, which can severely compromise the performance of regenerated active materials.[27, 28] Current prevention strategies are implemented through the optimization of storage and stripping conditions to minimize the content of aluminum and copper fragments and mitigate impurity effects. Alternatively, excessive aluminum and copper impurities can be selectively removed, and impurity thresholds can be explored to convert waste into useful materials while enhancing the performance of recovered products[29] (Figure 6.6).

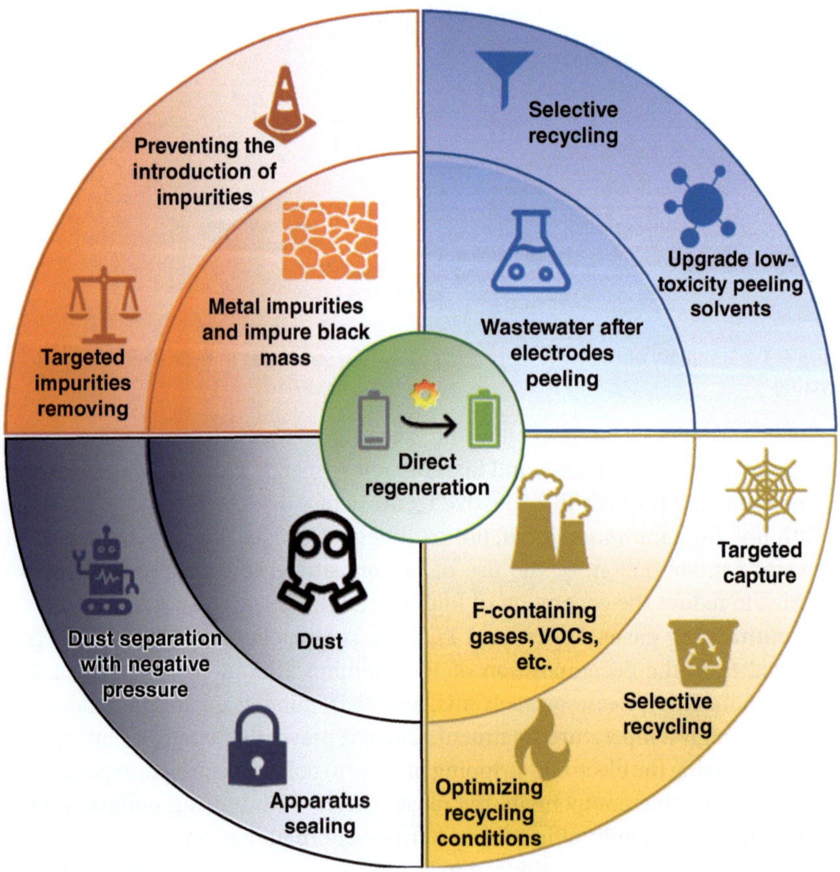

Figure 6.6 Summary of the pollutants and prevention strategies in direct regeneration recycling.

6.1.3 Comparative Analysis of Current Battery Recycling Technologies

To address the pollution generated during LIB recycling, cleaner recycling processes are being implemented by leading companies and research institutions to reduce pollutant emissions and enhance pollution prevention efficiency. In pyrometallurgical recycling, some companies have introduced oxygen-enriched smelting and waste heat recovery technologies. These processes reduce the amount of combustion flue gas and utilize the thermal energy of high-temperature gases for power generation or heating, thereby improving energy efficiency and indirectly lowering pollutant emissions. Some companies have also integrated lithium recovery into the pyrometallurgical process. For example, lithium can be fixed in specific phases by adding capture agents in the furnace, enabling subsequent extraction from slag and preventing lithium loss.

In hydrometallurgical recycling, advanced international practices employ a combined "pretreatment + mild hydrometallurgy" process. Batteries are first disassembled, and metals such as copper and aluminum are subsequently recovered through physical separation. The electrolyte and plastics are then removed through vacuum pyrolysis, and metals are subsequently leached using organic acids. The prior removal of most organic substances and impurities significantly reduces the volume of waste liquid generated in the subsequent hydrometallurgical stage. Organic acids can also be recovered via heating and evaporation, enabling the recycling of valuable reagents. This clean process effectively reduces acid consumption and wastewater generation, making it one of the more environmentally friendly solutions. Hydrometallurgical companies in China have achieved closed-loop recycling, with waste liquids generated during the process further concentrated and crystallized through multistage evaporation, as employed by BYD and CALB. The recovered water is reused in the leaching process, while the evaporative crystallization products are further processed to extract by-products such as Na_2SO_4, thereby generating additional revenue. This approach enables complete reuse of wastewater with zero discharge, thereby eliminating the risks associated with wastewater pollution.

In addition, although direct regeneration is still in the laboratory stage and its scalability for industrial applications requires further development, it generates far less pollution and fewer hazardous byproducts compared to conventional pyrometallurgical and hydrometallurgical recycling methods. Successful scaling of direct regeneration would provide a promising pathway toward environmentally sustainable battery recycling.[30] Overall, there is a clear trend of current technologies evolving toward cleaner production, with the realization of truly clean recycling expected in the future.

6.1.4 Demands for Battery Recycling

6.1.4.1 Demands for Battery Recycling Industry

Significant improvements in pollution prevention have been achieved under the guidance of leading companies; however, the industry still faces severe challenges. The main issues can be categorized into four categories: illegal recycling behaviors by unqualified workshops, poor matching of technical management ability with execution power, lack of risk supervision and treatment of new pollutants, and insufficient complete-chain control.

1. **Illegal recycling behaviors by unqualified workshops:** Numerous unregulated small companies employ primitive methods, such as open-air burning, manual disassembly, or simple acid leaching (e.g., direct soaking in H_2SO_4 or HCl), to recover only high-value metals like cobalt and nickel. Waste acids, containing heavy metals such as lead and cadmium, and residual organic solvents are often discharged without treatment. In provinces such as Guangdong and Jiangxi, soil contamination has been found to exceed standard limits by more than 10-fold, and high levels of cobalt and nickel have been detected in groundwater. The underlying causes are associated with the concealment and mobility of unregulated companies, which are often located on the urban–rural fringe or at the periphery of industrial parks with weak oversight. Enforcement capacity is considered insufficient, as in some regions only a few inspectors are assigned to oversee thousands of companies. Penalties remain low, frequently below 50,000 RMB, far less than the cost of pollution prevention, resulting in temporary crackdowns that fail to eliminate such operations. Data indicates that over 15% of degraded batteries are still processed through unregulated industrial chains.
2. **Poor matching of technical management ability with execution power:** Some companies reduce costs by bypassing wastewater treatment systems during peak operation or by diluting and discharging high-concentration wastewater, leading to serious emissions violations. In hydrometallurgical processes, uncontrolled VOC emissions may occur due to the degradation of extraction tanks and evaporators. Additionally, flue gas purification systems are often minimally automated, and pH instability in desulfurization towers can result in temporary SO_2 emissions exceeding regulatory limits. One of the main causes is that environmental protection facilities demand substantial capital investment (20–30% of total investment), yet small companies, with net profits generally below 5%, lack sufficient funding for continuous operation, resulting in a "compliance–loss" dilemma.
3. **Lack of risk supervision and treatment of new pollutants:** Emerging pollutant risks remain largely unregulated. For instance, N-methyl-2-pyrrolidone (NMP), a key solvent used in electrode coatings with high reproductive toxicity, can be released at 0.5–1.2 kg per ton of batteries during recycling, yet no emission standards have been established. Fluorinated organic decomposition products from electrolytes are both persistent and bioaccumulative, and have been detected at concentrations exceeding regulatory limits near recycling companies in North America. In hydrometallurgical processes, leaks of extractants such as TBP/kerosene result in highly concentrated organic wastewater, with chemical oxygen demand (COD) exceeding 100,000 mg/L. The main causes include insufficient removal of biochemical methods for NMP (<40%), the high cost of advanced oxidation technologies (e.g., O_3 catalysis, up to 80 RMB per

ton of water), the technical and economic challenges of degrading fluorinated organic compounds, and the slow and costly development of green aqueous extractants in a market still dominated by conventional organic extractants.

4. **Insufficient complete-chain control:** From 2021 to 2023, 23 fires involving LIB storage were reported globally, highlighting the frequent occurrence of accidents. One notable example is the explosion at a storage facility in Arizona, United States, where HF smoke was released and evacuation was enforced within a 10km radius. Several gaps in logistics and storage management have been identified. During collection, damaged batteries are transported by individual recyclers in ordinary trucks without short-circuit protection. During transport, leak emergency kits are lacking in 70% of companies, and drivers are generally not provided with professional training. In storage, many warehouses fail to meet fire separation requirements, and proper temperature and humidity monitoring are often absent.

To address these issues, four key areas of industry development are required.

1. **Strengthen "targeted supervision" and establish a multidimensional control system:** This involves establishing a full-lifecycle traceability platform and battery coding system to track production, sales, and recycling, while integrating blockchain to ensure data integrity and enable closed-loop monitoring. Intelligent identification and regulatory response should be enhanced by combining satellite remote sensing, drone-based infrared thermal imaging, and ground-based artificial intelligence (AI) cameras into a multisource detection and warning platform, with automatic transmission of coordinates to enforcement authorities for precise intervention.
2. **Innovate technical and economic models to handle the "cost-efficiency" contradiction:** Chemical and energy consumption can be reduced through modular wastewater treatment and upgraded treatment schemes, thereby addressing the constraints faced by small enterprises.[31] Real-time monitoring and adjustments are enabled through intelligent operation and maintenance supported by automated control systems, preventing regulatory exceedances. Emerging pollutant recovery, such as in situ NMP regeneration and advanced treatment of fluorinated compounds, should be prioritized in R&D investment, thereby mitigating pollutant impacts.[32]
3. **Upgrade the standard system and facilitate the new pollutants control:** Characteristic pollutant limits should be added in reference to international standards, and equipment requirements are being enforced for the recycling industry. For example, in hydrometallurgical recycling companies, sealed negative-pressure reactors with a TBP/kerosene system are required to be installed and equipped with leakage detection sensors to reduce fugitive emissions. In addition, mandatory upgrades of transportation and storage safety standards are being enforced, together with improvements in fire and explosion prevention measures. To further enhance accident resilience, explosion-resistant structures are being promoted, while intelligent linkage systems are being established to automatically treat toxic and hazardous gases in the event of safety incidents, thereby reducing the risk of secondary pollution.

4. **Establish a two-way "incentive-restraint" mechanism:** Establish a positive incentive by providing subsidies for battery recycling companies and offering R&D funding for core technology demonstrations. Implement a reverse constraint system by setting up a pollution disposal fund and imposing taxes on batteries to support the development of detection technologies for new pollutants and the restoration of contaminated sites, thereby feeding back into regulatory efforts and technological R&D to achieve low-carbon process innovations (Figure 6.7).

6.1.4.2 Demands for Battery Recycling Technologies

Based on the analysis of pollution risks associated with the recycling processes presented previously, it is evident that the three technological routes, pyrometallurgical recycling, hydrometallurgical recycling, and direct regeneration, each exhibit distinct advantages and limitations. They face different challenges in terms of engineering applicability and environmental performance. A comparison of their energy consumption, carbon emissions, and economic returns is provided in Figure 6.8.

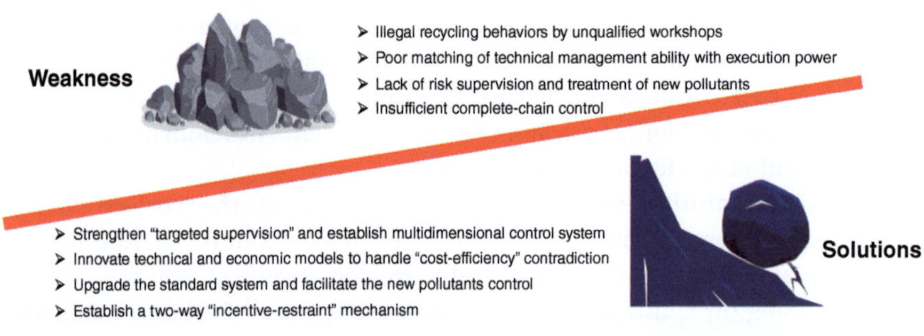

Figure 6.7 Challenges and demands of the degraded battery recycling industry.

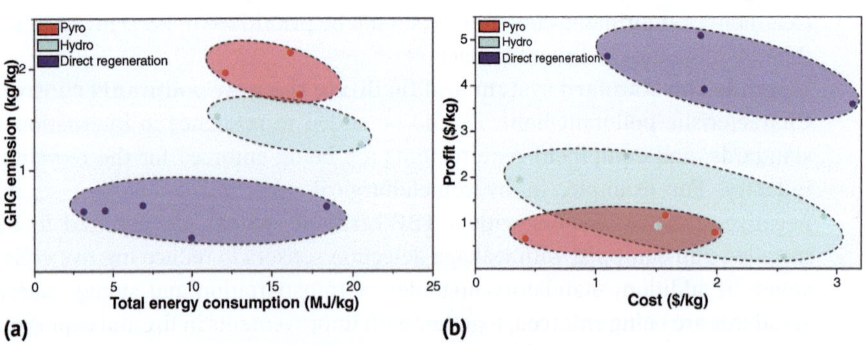

Figure 6.8 Comprehensive comparison of (a) energy consumption and carbon footprint and (b) economic performance for pyrometallurgical, hydrometallurgical, and direct regeneration recycling.

1. **Disadvantages of pyrometallurgical recycling:** The main limitations of pyrometallurgical recycling include high energy consumption, substantial emissions, and low lithium recovery rates. High-temperature processing requires large amounts of fuel or electricity, resulting in significant greenhouse gas emissions that largely offset the environmental benefits of recycling and reduce overall efficiency. Under the current context of energy transition and the "dual carbon" goals, these factors constrain the sustainability of pyrometallurgical recycling.

   In addition, pyrometallurgical processes face difficulties in recovering light metals such as lithium. Lithium tends to volatilize at high temperatures and become incorporated into slag or flue dust, preventing its direct recovery. Secondary processes are required to extract lithium, further increasing energy consumption while yielding limited recovery. In practice, most pyrometallurgical recycling plants directly use lithium-containing slag for building materials preparation, leading to resource loss and reduced efficiency.[33]

   Another disadvantage is the high technical and cost requirements for flue gas purification. During recycling, large amounts of fluorinated gases, VOCs, PCDDs, CO_x, NO_x, and dust are released, requiring multistage purification. Small and medium-sized recycling companies often lack the capacity to implement full-scale flue gas treatment, resulting in significant environmental risks. Moreover, the metal alloy products generated by pyrometallurgical processes typically require subsequent hydrometallurgical refining for further metal separation. Environmental drawbacks of pyrometallurgical recycling have prompted process improvements, such as lower operating temperatures, clean-energy furnaces, and flue gas heat recovery.
2. **Disadvantages of hydrometallurgical recycling:** Although hydrometallurgical recycling offers higher recovery rates and relatively lower energy consumption, it also presents notable environmental challenges. The process generates substantial amounts of wastewater and secondary residues, since each stage of leaching, extraction, and precipitation can produce high-concentration acidic or saline effluents. Without closed-loop recycling, the discharge of wastewater increases markedly,[34] and the effluents often contain complex ions, such as fluorine and phosphorus, that are difficult to remove. Furthermore, conventional neutralization leads to the formation of considerable quantities of filter sludges, including gypsum and metal hydroxide residues.

   High reagent consumption and numerous by-products are also observed in hydrometallurgical recycling. Extensive use of acids, bases, and extractants raises operational costs, and the environmental impact of producing these chemicals further diminishes process sustainability. If leachates and extractants are not recycled, the economic and environmental benefits are further reduced. Moreover, hydrometallurgical processes are typically optimized for specific chemical formulations, which restrict their flexibility. Processing mixed wastes with diverse compositions requires pre-classification; otherwise, element interference may lower recovery efficiency or result in incomplete precipitation.[35]

Mechanical crushing and classification steps also introduce secondary issues such as dust and noise emissions.

Safety risks are associated with the handling of multiple chemicals, including corrosive acids, alkalis, and flammable substances. Prolonged process durations, complex equipment requirements, and high capital and operational expenditures are considered additional barriers for small- and medium-sized recycling enterprises. As a result, environmental protection measures are often simplified to reduce costs, which may ultimately lead to severe environmental pollution incidents. Potential improvements in hydrometallurgical recycling have been proposed, including process optimization to reduce wastewater and residue generation, minimization of high-concentration acid consumption, utilization of novel aqueous extractants to mitigate pollution, enhanced recycling of reagents, and simplification of recovery procedures.[36]

3. **Disadvantages of direct regeneration recycling:** Although direct regeneration recycling has demonstrated significant environmental advantages at the laboratory scale, including low reagent consumption, low energy use, and minimal pollutant generation, current technologies remain immature and face multiple obstacles to industrial application.[4] Firstly, direct regeneration recycling requires battery materials to be separated from impurities. Different cathode materials are typically processed individually, while mixed powders are difficult to recover efficiently. This requirement necessitates precise disassembly and sorting at the initial stage, thereby increasing production complexity and associated costs. Secondly, the performance of regenerated materials must be thoroughly evaluated. Although material recovery has been successfully demonstrated at the gram scale, consistency and impurity thresholds for batch regeneration remain uncertain, thereby posing challenges for downstream manufacturers and constraining market acceptance.[37] Additionally, full-component battery recycling is constrained in direct regeneration processes, which primarily focus on cathode materials and lack effective strategies for anode recovery.

Strict process parameter control is required in direct regeneration recycling, including precise control of material properties, reaction conditions, and heating profiles, which challenges reproducibility at the industrial scale. During each cycle, impurities (e.g., residual conductive agents, current collectors, or electrolytes) can be accumulated, potentially affecting battery performance and safety,[38–40] and mechanisms for their removal or suppression must be implemented. From a commercial perspective, integration with downstream cell manufacturers and automakers is necessary to ensure that recovery and reuse are conducted in a timely manner; otherwise, materials may be subjected to degradation, impurities may be accumulated, and repairability may be compromised.

Direct regeneration recycling is characterized by clear limitations in scalable recovery and batch adaptability, making industrial operation and the processing of diverse degraded batteries challenging, and preventing the short-term replacement of conventional pyrometallurgical and hydrometallurgical processes.

In summary, pyrometallurgical recycling is hindered by high energy consumption, difficulties in controlling flue gas and slag pollution, and low recovery rates of some high-value components. Hydrometallurgical recycling is constrained by challenges in wastewater treatment, low reagent efficiency, and limited recycling of materials and chemicals. Direct regeneration recycling is restricted by batch stability and compatibility issues at the industrial scale. To address these challenges, the main development needs for recycling processes are as follows:

1. **Development and upgrading of clean recycling processes:** Under the "dual carbon" goals and the concept of a circular economy, the development and upgrading of clean and efficient degraded battery recycling processes are urgently required.

For pyrometallurgical recycling, cleaner heat sources should be introduced, and low-temperature molten salt electrolysis can be developed to partially replace existing high-temperature methods, thereby enabling the provision of zero-carbon heat. Examples include the use of concentrated solar thermal energy,[41] green electricity, or hydrogen-fueled smelting furnaces to reduce carbon emissions. Residual heat should be fully recovered to improve thermal efficiency, and clean furnaces upgraded to ultrahigh temperatures with short heating durations can be employed to reduce energy consumption and ensure complete decomposition of PCDDs. Volatile elements, such as lithium, should be selectively enriched and captured. By integrating multiple technical solutions, graded purification and automated directional recovery can be realized, while the replacement of slag-based hydrometallurgical refining with short-process refining allows acid wastewater to be avoided and metal purity to be improved.

For hydrometallurgical recycling, optimized closed-loop processes should be designed to achieve near-zero emissions. Wastewater should be controlled at the source through selective leaching, separation, concentration, and electrodeposition to reduce chemical consumption. The use of alkaline precipitants should be avoided, significantly reducing solid waste and effluent.[42, 43] Solid residues should be recovered or reused; for instance, filter residues can be employed in other industries. Recovery processes should be simplified, with intelligent, quality-based separation and real-time leaching monitoring implemented. Reagent dosages can be dynamically adjusted using AI,[44] and high-value elements can be efficiently recovered selectively. Microchannel reactors can replace traditional stirred tanks to improve mass transfer, reduce reagent use, and shorten reaction times.

For direct regeneration recycling, focus should be placed on precise disassembly and sorting, standardized and stable regeneration indices and processes, cost-effective high-efficiency regeneration technologies, intelligent process control, and innovative business models. Advanced technologies such as AI, multispectral sorting, and low-temperature dissociation can rapidly identify and process materials, solving compatibility issues. Standardized indicators and processes are established to improve batch stability. Low-cost, efficient regeneration technologies, multi-material simultaneous regeneration, and homogenized recovery are applied to reduce the need for frequent process adjustments.[45–47] Digital twins, AI, automation, and machine learning can be used to dynamically optimize regeneration, monitor and remove impurities online, and ensure uniform performance and batch stability. Blockchain-based full-lifecycle traceability can be implemented to guarantee low impurity content and enable dynamic process adjustments. Regional "receive-and-regeneration" centers can be established to facilitate efficient disassembly, regeneration, and recovery, while rapid failure mechanism identification and targeted regeneration are employed to maximize recovery rates.[48, 49]

2. **Reducing secondary pollution risks and promoting green, safe development:** Secondary pollution must be minimized to achieve green and safe development. Large amounts of electrolytes and PVDF binders are present in pyrometallurgical, hydrometallurgical, and direct regeneration processes. Traditional flue gas absorption towers are inefficient, and equipment failures may lead to toxic gas leaks. High-efficiency composite or low-cost solvents are required to selectively remove and decompose hazardous gases. Acid mist and flue gas from pyrometallurgical and hydrometallurgical processes require similar treatment strategies.

For lithium volatilization in pyrometallurgical processes, $CaCl_2$ is typically added to capture the lithium element and convert it to LiCl, but Cl-containing flue gas releases pose environmental hazards. Novel recovery methods, such as molten salt electrolysis,[50] can avoid flue gas generation while improving lithium and metal recovery. In hydrometallurgical recycling, the use of traditional organic extractants should be minimized, and aqueous electrochemical deposition techniques should be developed to fully utilize the residual energy of retired batteries for electrolysis. Metals can be directly reduced and deposited from the aqueous solution, eliminating the need for organic extractants and preventing the potential for extractant leakage and water pollution.[51] Such approaches to minimizing secondary pollution are consistent with the principle of clean production, as they reduce emissions at the source.

3. **Improving resource recovery and economic efficiency:** By reducing waste emissions, more material can be reintroduced into the industrial cycle, thereby maximizing resource utilization. Former waste residues can be converted into marketable products, allowing solid waste to be eliminated while generating revenue and establishing a positive feedback loop.[52] Higher resource recovery rates decrease reliance on primary mining, mitigating environmental impacts and providing economic benefits. In addition, clean recycling processes can reduce treatment and discharge costs, lower environmental penalties and carbon taxes, and increase profits through the utilization of by-products (Figure 6.9).

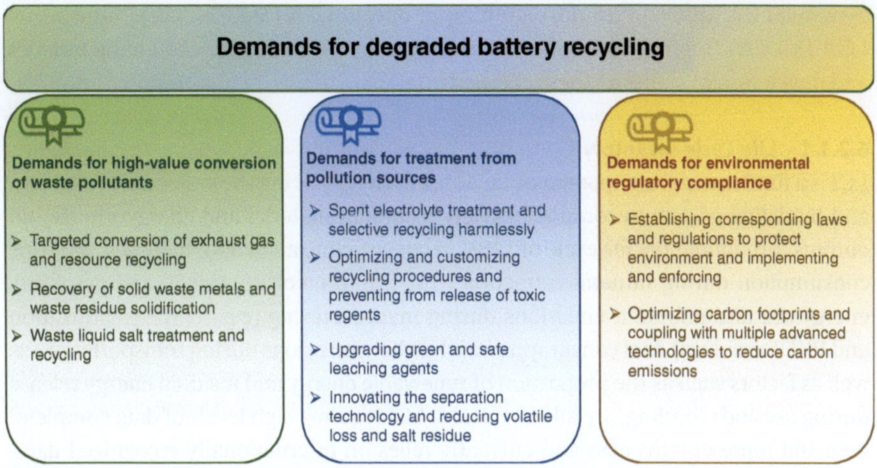

Figure 6.9 Summary of demands for degraded battery recycling technologies.

6.2 Life Cycle Assessment

Although LIBs are considered a clean energy storage technology, carbon emissions are generated at each stage of their life cycle, including manufacturing, transportation, use, and recycling. This is particularly evident in the recycling and reuse of degraded batteries, where substantial carbon emissions and other pollutants are produced during recovery. In both pyrometallurgical and hydrometallurgical recycling, recovered materials are often ultimately processed into high-value battery products to improve economic returns, and the subsequent manufacturing and transportation stages inevitably generate additional carbon emissions. Life cycle assessment (LCA) is a standardized and scientific methodology for evaluating the environmental impacts of products or services. It has been widely applied to quantify carbon emissions and other environmental burdens across the manufacturing, use, and recycling stages of LIBs, enabling the identification of key stages for process improvement. The implementation of LCA can effectively facilitate the development of LIB recycling technologies, optimize resource utilization, and reduce environmental pollution.[53–55]

6.2.1 Current Status of LCA for LIBs

In the LCA of LIBs, multiple evaluation techniques are applied, covering the entire process from resource acquisition and material preparation to battery manufacturing, use, end-of-life treatment, and recycling. These methods are used not only to assess environmental impacts but also to quantify energy consumption, carbon emissions, resource depletion, toxicity risks, and other relevant indicators. Currently, the most widely applied and representative methods are classified into five categories: life cycle inventory (LCI),[56] life cycle impact assessment (LCIA),[57] dynamic LCA (dLCA),[58] multi-region input-output LCA (MRIO-LCA),[59] and social LCA (S-LCA) (Figure 6.10). The following sections summarize the characteristics and development status of these methods.

6.2.1.1 Life Cycle Inventory

LCI is a fundamental component of LCA, primarily involving the systematic collection and modeling of data to comprehensively record all material and energy inputs and outputs throughout the life cycle of LIBs. In constructing an LCI for batteries, resource consumption during mineral extraction (e.g., lithium, cobalt, nickel, and graphite), energy use and pollutant emissions during manufacturing (e.g., NMP volatilization and PVDF residues), fuel consumption and carbon emissions during transportation, as well as factors such as the proportion of renewable energy and residual energy release during use and recycling, are all considered. LCI requires high levels of data completeness and representativeness and currently relies on internationally recognized databases (such as Ecoinvent, GREET, GaBi, and ELCD) as well as enterprise-measured data, gradually establishing a dedicated inventory system for the LIBs industry.[60]

With the widespread adoption of sensors and Industrial Internet of Things (IIoT) technologies, the efficiency and accuracy of data collection have been significantly improved, and LCI is evolving from static, coarse material flow statistics to high-precision, dynamic data integration.[61] Historically, LIBs and LCIs have relied on international databases such as Ecoinvent and GREET, but most of these data originate from the EU and the US, which do not fully represent Chinese lithium extraction processes, battery manufacturing routes, and energy consumption structures, resulting in limited applicability. In recent years, driven by the "dual carbon" policy, efforts have been made in China to construct localized LCI databases, such as the Chinese Life Cycle Database (CLCD) and regional LIBs inventory models. Moreover, the process data real-time acquisition technology based on

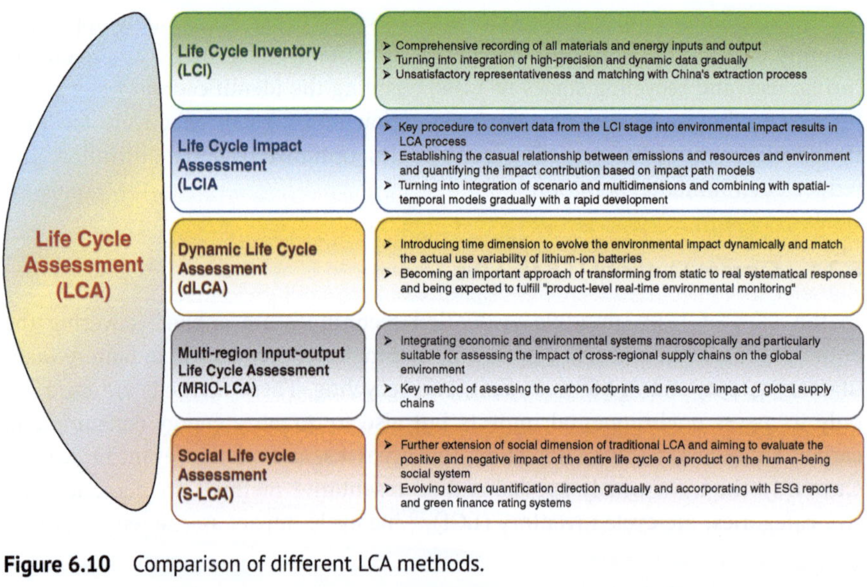

Figure 6.10 Comparison of different LCA methods.

enterprise-level manufacturing execution systems (MES), enterprise resource planning (ERP) platforms, and smart manufacturing platforms makes dynamic product LCI modeling possible, providing high-frequency, high-resolution inputs for tracking product carbon footprints and environmental impacts. From a research perspective, open-source modeling tools (e.g., openLCA and Brightway2) are widely applied, supporting various boundary settings and scenario assumptions. Overall, the development trend of LCI is toward localization, dynamic modeling, and modularization. The quality and accuracy of LCI directly determine the scientific validity and reliability of the entire LCA system. It has become the core of data-driven modeling, underpinning all fundamental processes of environmental impact modeling and providing high-precision inputs for downstream impact assessment.

6.2.1.2 Life Cycle Impact Assessment

LCIA is a critical step in the LCA, used to convert data from the LCI stage into environmental impact results. Its core methodology is based on impact pathway modeling, which establishes causal relationships between various substance emissions (e.g., CO_2, NO_x, SO_2, heavy metals, and particulates) and resource consumption (e.g., freshwater, land, and minerals) with corresponding environmental impacts (e.g., global warming, ozone depletion, acidification, eutrophication, and ecotoxicity), allowing the contribution of each life cycle stage to be quantified.[62] As for LIBs, LCIA has gradually developed industry-specific impact factor sets, such as "resource scarcity indicators" for battery-critical metals (e.g., cobalt, lithium, and nickel) and "water toxicity indicators" to account for electrolyte leakage effects on aquatic ecosystems.

Commonly used LCIA methods include ReCiPe, TRACI, CML, Impact 2002+, and ILCD. These methods differ in impact pathway construction depending on geographic region and temporal scale.[63] The development of LCIA is shifting from standardized indicators toward scenario-specific refinement and multidimensional integration. Traditional models, such as ReCiPe, TRACI, and CML, provide broad impact categories, including global warming, acidification, eutrophication, and ecotoxicity, suitable for overall assessment of LIBs production and recycling. However, the specificity of the LIBs industry, such as the use of rare metals, energy-intensive manufacturing, and complex waste characteristics, means that existing methods cannot fully capture its unique environmental burdens.

To address this limitation, regionalization and customization of LCIA are being promoted. Examples include the introduction of weighted depletion factors for lithium, cobalt, and nickel scarcity, as well as modifications to water toxicity models to reflect the effects of specific solvents and additives. LCIA is also being refined to higher spatial and temporal resolution, incorporating localized adjustments for energy mix, land use, and pollutant emission indicators, particularly in China. New-generation tools, such as LC-IMPACT and ILCD 2011, have introduced geographic information system (GIS)-based impact pathway simulation and ecosystem service loss assessment, enabling evaluation of environmental transfers between production regions (e.g., China and South America) and consumption regions (e.g., the EU). Emerging multifactor coupled assessment models are also being developed to consider synergies between battery recycling and energy feedback. In the future, LCIA is expected to extend beyond environmental impacts to integrate broader sustainability dimensions, including ecosystem service losses and social acceptance, thereby supporting multi-scale data integration, cross-country comparison, and sustainable design feedback.

6.2.1.3 Dynamic LCA

dLCA differs from traditional LCA methods, which are mostly based on static modeling and often neglect the temporal influence on environmental impacts. By incorporating the time dimension, dLCA allows environmental impacts to evolve dynamically across the entire life cycle, more accurately reflecting variability in actual LIB usage. For instance, during the use phase, a battery may undergo capacity degradation, experience changes in the grid mix (e.g., gradual transition to cleaner energy), and be subjected to varying charge-discharge patterns, all of which significantly influence carbon footprint and energy efficiency. dLCA tracks impacts across multiple stages under dynamic boundaries using time-series simulation and system dynamics modeling.

Currently, dLCA technology is advancing rapidly and is becoming an essential tool for transforming LCA from static evaluation to real-system response assessment. It is frequently integrated with GIS, big data analytics platforms, and real-time grid data (e.g., time-varying carbon factors), making it suitable for environmental management of large-scale electric vehicle fleets and ESS. In the field of LIB, dLCA is particularly useful for evaluating second-life applications, such as cascade utilization, enabling dynamic comparisons of total impacts between direct recycling and recycling after second-life use. This approach supports policy development, technology pathway optimization, and strategic decision-making. With continued progress in cloud computing and AI-based modeling platforms, dLCA is expected to enable product-level, real-time environmental monitoring, facilitating design optimization and dynamic supply chain management, and becoming a core component of carbon accounting systems for battery companies.

6.2.1.4 Multi-region Input-output LCA

MRIO-LCA is a macro-level LCA approach that integrates economic and environmental systems, making it particularly suitable for evaluating the global environmental impacts of LIB supply chains across regions. In MRIO-LCA, the resources consumed and environmental burdens generated during battery production in a given region are allocated to its upstream global supply chain (e.g., lithium mining in South America, cobalt mining in the Democratic Republic of the Congo, separator production in Germany), thereby revealing the phenomenon of environmental transfer. The technical foundation of MRIO-LCA lies in MRIO tables (such as EXIOBASE, WIOD, and GTAP datasets), which, when combined with environmental satellite accounts, allow environmental factors to be assigned to each economic activity, enabling global tracking of environmental footprints. For battery manufacturers, MRIO-LCA can identify upstream stages and countries responsible for the highest environmental burdens, supporting the optimization of raw material procurement strategies and reducing environmental transfer effects.

At present, MRIO-LCA has evolved from a traditional environmental–economic research tool into a key approach for assessing the carbon footprint and resource impacts of global supply chains. It has developed rapidly in the context of globalization, and especially after its integration with global value chain (GVC) analysis, it provides strong guidance for policymaking. For instance, the EU Carbon Border Adjustment Mechanism (CBAM) has already adopted MRIO models as a technical basis for battery carbon footprint accounting. In the future, MRIO-LCA is expected to be applied more widely in regional comparative analyses, carbon border regulation, and environmental trade assessments, establishing itself as a crucial tool for macro-scale green policy development.

6.2.1.5 Social LCA

S-LCA extends traditional LCA to the social dimension, enabling the evaluation of both positive and negative impacts of a product life cycle on human social systems. Unlike environmental LCA, S-LCA does not focus on CO_2 emissions or energy consumption; rather, it addresses aspects such as labor rights, occupational health and safety, gender equality, community development, and business ethics. In the LIB supply chain, socially sensitive issues include child labor in cobalt mines in the Democratic Republic of the Congo, competition for water resources in South American salt lakes, occupational health risks during battery manufacturing, and the treatment of informal workers in degraded battery recycling. S-LCA is conducted using a comprehensive social indicator framework, incorporating sources such as the United Nations (UN) Sustainable Development Goals (SDGs), the UN Global Compact (UNGC), and the Organization for Economic Co-operation and Development (OECD) Guidelines, together with surveys, interviews, expert judgment, and statistical modeling.

Currently, advanced S-LCA approaches are guided by the UN Environment Programme-Society of Environmental Toxicology and Chemistry (UNEP-SETAC) S-LCA guidelines and supported by modeling systems such as the PSILCA database. Although S-LCA has not attained the level of universal standardization established in environmental LCA, it is increasingly applied in the field of battery, driven by social ethics pressures and cross-border procurement compliance requirements. The development trend is toward quantitative evaluation, with S-LCA being progressively integrated into corporate environmental, social, and governance (ESG) reporting and green finance rating systems. This integration promotes corporate social responsibility and enhances transparency and fairness across global battery supply chains.

6.2.2 Comparison of Different LCA Methods

The mainstream tools currently used in LCA include SimaPro, openLCA, Brightway2, GREET, PSILCA, GaBi, and eFootprint. These tools differ in their methodological frameworks and model construction approaches, and they are employed to evaluate the complete life cycle of LIBs. A detailed comparison of their respective advantages and limitations is provided.

6.2.2.1 SimaPro

SimaPro serves as a benchmark platform in academic research. It follows ISO 14040/14044 methodology, applies Monte Carlo-based multilevel uncertainty analysis, and integrates twelve global databases, including Ecoinvent and CLCD. It allows cradle-to-grave analysis across seven stages—raw material extraction, material processing, cell manufacturing, pack assembly, transportation, use, and recycling, and incorporates multiple LCIA methods such as ReCiPe, ILCD, TRACI, and IPCC.

Its core advantages are: (1) the most extensive international databases, such as Ecoinvent, Agri-footprint, and USLCI, covering key processes including lithium resources, electrolytes, and manufacturing, with multi-database integration and automatic selection of the optimal dataset; (2) support for multiple LCIA methods and social impact extensions, including social hotspot databases; (3) a mature graphical modeling interface, facilitating structural and sensitivity analyses; (4) capability for multi-scenario comparison, scenario simulation, and attributional analysis, suitable for policy and pathway decision-making; (5) built-in hierarchical modeling of pyrometallurgical and hydrometallurgical recycling processes, allowing adjustable recycled material replacement rates (0–100%) and real-time calculation of carbon offset benefits.

However, some limitations and challenges remain, including: (1) limited support for dLCA and MRIO-LCA, with weak temporal simulation capabilities; (2) insufficient localization of regional indicators, with significant gaps in Chinese data, leading to considerable differences in carbon footprint assessment and necessitating additional adjustments for the Chinese data; (3) oversimplification of recycling processes, such as the degradation of extractants in hydrometallurgical recycling (e.g., TBP decomposing to dibutyl phosphate) not being included in toxicity assessments, thereby overlooking potential risks to aquatic ecosystem.[64]

6.2.2.2 openLCA

openLCA is widely applied in academic research, enterprise pilots, and government policy evaluation. It supports customized model building and script development, and is compatible with multiple database formats, such as Ecoinvent, LCI datasets, PSILCA, and openLCA Nexus.

Its core advantages are as follows: (1) it is open-source with high scalability, making it suitable for advanced development and localized modeling by researchers; (2) it supports S-LCA modeling and allows the import of social databases such as PSILCA; (3) it has built-in advanced features, including sensitivity analysis, parameterized modeling, and graphical outputs.

However, certain limitations and challenges remain: (1) it has limited support for complex processes, such as multiple life cycles and multiregional models; (2) it lacks official technical support, relying mainly on community forums and documentation for problem solving.[65]

6.2.2.3 Brightway2

Brightway2 does not provide a graphical user interface, but it performs well in dLCA, sensitivity analysis, and large-scale batch simulations.

Its core advantages are as follows: (1) it offers strong programming capability, making it suitable for constructing customized evaluation models and dynamic life cycle simulations; (2) it can be integrated with scientific computing libraries such as NumPy and Pandas, as well as AI and machine learning models; (3) it has high adaptability, applicable to complex studies such as dLCA and recycling process evaluation; (4) it supports Brightway extensions, enabling GIS-based modeling and spatial analysis.

Nevertheless, some limitations and challenges remain: (1) its modeling process is cumbersome and highly unfriendly to beginners; (2) it lacks intuitive interaction support for policymakers and senior corporate decision-makers.[66]

6.2.2.4 GREET Model

The GREET model is primarily used for carbon footprint accounting of energy, transportation, and battery products, and it is widely applied in the fields of electric vehicles and energy storage batteries.

Its core advantages are as follows: (1) it focuses on energy and transportation systems, providing accurate modeling of energy consumption and carbon emissions of LIBs; (2) it contains extensive datasets on LIBs, hydrogen, and power systems (applicable to NMC, LFP, and LCO batteries); (3) it can be integrated with the US LCA database (USLCI), supporting comparative analysis of carbon reduction pathways.

Nevertheless, certain limitations and challenges remain: (1) it is mainly based on the US energy mix, and substantial modifications are required when applied to cases of China or EU; (2) it concentrates on greenhouse gases, with limited coverage of multidimensional impacts such as ecotoxicity and resource depletion; (3) it does not support social impact modeling and is therefore not suitable for S-LCA analysis.[67]

6.2.2.5 PSILCA

PSILCA can be deeply integrated with openLCA for evaluating social risks and impacts across the product life cycle, covering dimensions such as labor rights, occupational safety, gender equality, and community welfare.

Its core advantages are as follows: (1) it has broad data coverage, including social performance factors for more than 150 countries and multiple industries worldwide; (2) it is closely integrated with openLCA, enabling rapid construction of S-LCA models; (3) it supports scoring systems, risk ratings, and value-oriented indicators, facilitating integration with corporate ESG frameworks; (4) it provides strong differentiation capacity for high-risk sectors such as lithium and cobalt mining.

Nevertheless, certain limitations and challenges remain: (1) its data are primarily based on national or sectoral levels, lacking micro-level enterprise details; (2) its modeling complexity is relatively high, and interpretation of results requires advanced expertise; (3) some datasets have long update cycles, lagging behind real-world developments; (4) the lack of a unified methodology makes it difficult to directly integrate results with environmental LCA.[68]

6.2.2.6 GaBi

GaBi adopts a process matrix model, decomposing the battery life cycle into more than 500 unit processes. Each process includes an input matrix (energy, raw materials, and equipment), an output matrix (products, emissions, and waste), and an environmental flow matrix (18 indicators, including CO_2, SO_2, and COD). It supports LCI and LCIA modeling, scenario analysis, corporate carbon management, and is compatible with mainstream methodologies such as ReCiPe, CML, ILCD, and IPCC.

Its core advantages are as follows: (1) its industry database is authoritative and highly detailed, covering advanced European processes and material flows; (2) it has strong adaptability, supporting multiproduct systems, comparative scenarios, and policy simulation; (3) it is well-established in corporate applications, commonly used for compliance reporting under ISO 14040 and EN 15804.

Nevertheless, certain limitations and challenges remain: (1) its databases are mainly EU and US, and have not incorporated China battery traceability platform, leading to assumptions such as an average retirement battery transport distance of 800 km nationwide, whereas in practice the regional center model requires only 200 km, underestimating carbon emissions by about 60%; (2) it provides limited support for dLCA and struggles to handle time-series variables; (3) its recycling module is outdated, the wastewater treatment model for hydrometallurgical recycling still relies on conventional lime neutralization (producing 2.5 tons of gypsum slag per ton of batteries), without integrating advanced reverse osmosis membrane technology (which can reduce solid waste by 70%).[69]

6.2.2.7 eFootprint

eFootprint is a localized LCA and carbon footprint accounting software developed in China, designed for corporate carbon management, green product design, and life cycle environmental impact assessment under the national "dual carbon" strategy. The platform adopts a modular process design and supports LCI and LCIA model construction, product carbon footprint accounting, and alignment with national policies such as green manufacturing, energy efficiency standards, and carbon footprint labeling.

Its core advantages are as follows: (1) its datasets are highly localized, reflecting the actual conditions of China's energy, transportation, and manufacturing sectors; (2) it has built-in standard models for batteries such as NMC and LFP, enabling rapid modeling and result output; (3) it is aligned with national policy platforms and green factory standards, facilitating compliance management; (4) it is deployed in a Software-as-a-Service (SaaS) mode, lowering the entry barrier and supporting collaboration and data sharing.

Nevertheless, certain limitations and challenges remain: (1) its international influence and methodological coverage are relatively limited; (2) the openness of its data sources and transparency of model algorithms need improvement; (3) it has weak support for complex product systems, such as complete vehicles and MRIO scenarios; (4) advanced analysis functions, such as dynamic simulation and AI-assisted optimization, are still immature.

6.2.3 Development Trends of LCA Methods

Existing life cycle monitoring technologies have achieved significant progress; however, a series of technical challenges remain in practical applications. Analysis and comparison of different methods indicate several key development directions.

6.2.3.1 Data Integration and Standardization

With the expansion of the global battery industry chain, data integration and standardization have become increasingly critical. Different countries and regions currently employ varied standards, and the absence of unified technical criteria presents challenges for multinational companies and international collaboration. In the future, the International Organization for Standardization (ISO) and global battery recycling associations should be leveraged to strengthen cross-border coordination of data and standards, and to promote the establishment of unified protocols for life cycle monitoring.

6.2.3.2 Intelligentization and Automation

Life cycle monitoring technologies are expected to advance toward higher levels of intelligentization and automation. Through the application of the IoT, big data analytics, and AI, automatic monitoring and intelligent optimization can be achieved across all stages of battery production, use, and recycling. In particular, during the battery recycling and reuse stage, the deployment of intelligent equipment is expected to significantly enhance recovery efficiency and resource utilization.

6.2.3.3 Life Cycle Optimization and Closed-loop Economy Development

Future life cycle monitoring should extend beyond the assessment of environmental impacts to support the optimization of the entire battery life cycle. By analyzing energy consumption, material flow, and waste generation at all stages, battery design, production, and recycling processes can be optimized. This approach will facilitate the development of a closed-loop economy for batteries, enabling efficient resource use while minimizing environmental pollution.

6.2.3.4 Cross-field Technology Collaboration

Life cycle monitoring is no longer solely an environmental engineering application; future development will require deeper collaboration across disciplines. As for battery recycling, expertise from materials science, chemical engineering, mechanical engineering, data science, and related fields must be combined to improve recovery efficiency, reduce costs, and minimize environmental impacts (Figure 6.11).

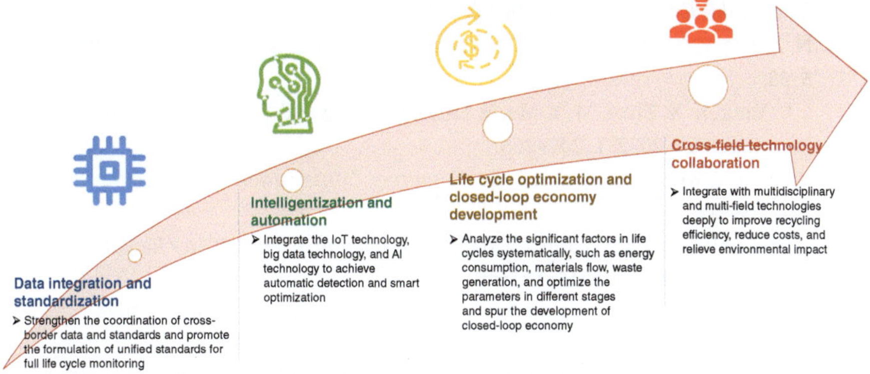

Figure 6.11 Development trends of LCA for degraded LIBs recycling.

6.3 Conclusions

To establish a clean, efficient, and closed-loop recycling system for LIBs, it is essential to significantly strengthen pollution prevention, thereby minimizing the negative environmental impacts of degraded batteries and achieving green and sustainable development throughout the entire life cycle of LIBs. Specifically, efforts should be made to achieve zero harmful emissions and zero environmental incidents during the recycling process of degraded LIBs, while reaching a recycling efficiency close to 100%. In addition, LCA should be applied to optimize processes, ensuring that the carbon footprint per kilogram of recycled LIB materials is significantly reduced compared with current benchmarks. This would provide strong support for the sustainable development of the global new energy industry and help build a recycling ecosystem for degraded batteries that combine economic feasibility with environmental friendliness. In such a system, valuable materials such as lithium, cobalt, and nickel would be reused, primary resource consumption and pollutant emissions would be minimized, and the principles of "resource circulation, environmental sustainability, and benefits for all" would be fulfilled.

References

1 J. Wang, J. Ma, Z. Zhuang, Z. Liang, K. Jia, G. Ji, G. Zhou and H.-M. Cheng, *Chemical Reviews*, 2024, **124**, 2839–2887.

2 Z. Lai, J. Long, Y. Lu, F. Luo, L. Zeng, W. Lai, Y. Li, Q. Qian, Q. Chen, K. Zhang, Z. Yan and J. Chen, *Advanced Energy Materials*, 2025, **15**, 2501009.

3 B. K. Biswal, B. Zhang, P. T. M. Tran, J. Zhang and R. Balasubramanian, *Chemical Society Reviews*, 2024, **53**, 5552–5592.

4 K. Jia, G. Yang, Y. He, Z. Cao, J. Gao, H. Zhao, Z. Piao, J. Wang, A. M. Abdelkader, Z. Liang, R. V. Kumar, G. Zhou, S. Ding and K. Xi, *Advanced Materials*, 2024, **36**, 2313273.

5 A. Cornelio, A. Zanoletti and E. Bontempi, *Current Opinion in Green and Sustainable Chemistry*, 2024, **46**, 100881.

6 N. Zhang, Z. Xu, W. Deng and X. Wang, *Electrochemical Energy Reviews*, 2022, **5**, 33.

7 J. L. Yang, X. X. Zhao, M. Y. Ma, Y. Liu, J. P. Zhang and X. L. Wu, *Carbon Neutralization*, 2022, **1**, 247–266.

8 Y. Zhao, Y. Wen, Y. Yang and S. Xu, *Resources, Conservation and Recycling*, 2025, **221**, 108403.

9 T. Miyashita, K. Yasuda and T. Uda, *Environmental Science: Water Research & Technology*, 2024, **10**, 1245–1255.

10 J. Xiao, W. Sun, Y. Song, Q. Li, M. Wu and Y. Yang, *Hydrometallurgy*, 2025, **234**, 106471.

11 H. Zhang, L. Zhu, Y. Zhang, P. Héroux, L. Cai and Y. Liu, *Water Research*, 2024, **253**, 121316.

12 H. Zhang, Z. Wang, L. Wei, Y. Liu, H. Dai and J. Deng, *Chinese Journal of Catalysis*, 2024, **61**, 71–96.

13 Y. Guo, Q. Guan, X. Li, M. Zhao, N. Li, Z. Zhang, G. Fei and T. Yan, *Energy & Environmental Science*, 2025, **18**, 5539–5551.

14 A. Holzer, L. Wiszniewski, S. Windisch-Kern and H. Raupenstrauch, *Metals*, 2022, **12**, 1642.

15 J. Liu, Z. Wang, Z. Zhang, Y. Qi and Z. Guo, *Chemical Engineering Journal*, 2025, **509**, 161493.

16 X. Zhang and M. Zhu, *Green Chemistry*, 2024, **26**, 7656.

17 J. Wu, L. Xiao, L. Shen, J.-J. Ran, H. Zhong, Y.-R. Zhu and H. Chen, *Rare Metals*, 2023, **43**, 879–899.

18 A. Saidi, R. El Khawaja and D. C. Boffito, *ACS Engineering Au*, 2023, **4**, 49–70.

19 W. Wang, S. Aleid, Y. Shi, C. Zhang, R. Li, M. Wu, S. Zhuo and P. Wang, *Joule*, 2021, **5**, 1873–1887.

20 M. d. M. Cerrillo-Gonzalez, J. M. Paz-Garcia, M. Muñoz-Espinosa, J. M. Rodriguez-Maroto and M. Villen-Guzman, *Journal of Power Sources*, 2024, **592**, 233870.

21 A. D. A. B. A. Sofian, S. R. Majid, K. Kang, J.-K. Kim and P. L. Show, *Progress in Materials Science*, 2025, **153**, 101478.

22 S. Natarajan and S. Noda, *Materials Science and Engineering: R: Reports*, 2025, **164**, 100976.

23 Y. Hua, Z. Xu, B. Zhao and Z. Zhang, *Green Chemistry*, 2022, **24**, 8131–8141.

24 Z. Chen, R. Feng, W. Wang, S. Tu, Y. Hu, X. Wang, R. Zhan, J. Wang, J. Zhao, S. Liu, L. Fu and Y. Sun, *Nature Communications*, 2023, **14**, 4648.

25 H. Lei, X. Cui, J. Li, Z. Zeng, C. Zhu, X. Ji, W. Sun, Y. Yang and P. Ge, *Energy Storage Materials*, 2025, **77**, 104201.

26 T. W. Hsu, A. Lipson and Z. Zhang, *Advanced Energy Materials*, 2025, **15**, 2500655.

27 C. Xing, H. Da, P. Yang, J. Huang, M. Gan, J. Zhou, Y. Li, H. Zhang, B. Ge and L. Fei, *ACS Nano*, 2023, **17**, 3194–3203.

28 Z. Zeng, H. Lei, J. Li, B. Wang, S. Lei, X. Ji, W. Sun, Y. Yang and P. Ge, *Chemical Engineering Journal*, 2024, **499**, 155616.

29 B. Wang, Y. Dong, Z. Zeng, H. Lei, Y. Yang and P. Ge, *Journal of Power Sources*, 2024, **621**, 235299.

30 Z. Yuan, Y. Dong, Z. Zeng, W. Zhao, B. Wang, H. Lei, W. Sun, X. Ji, Y. Yang and P. Ge, *Chemical Engineering Journal*, 2024, **491**, 151948.

31 J. Zhang and Q. Ge, *Water Research*, 2024, **260**, 121939.

32 锂离子电池用再生 N-甲基-2-吡咯烷酮, T/CIESC 85-2025.

33 H. Rostami, J. Valio, P. Tynjälä, U. Lassi and P. Suominen, *ChemPhysChem*, 2024, **25**, e202400459.

34 G. Harper, R. Sommerville, E. Kendrick, L. Driscoll, P. Slater, R. Stolkin, A. Walton, P. Christensen, O. Heidrich, S. Lambert, A. Abbott, K. Ryder, L. Gaines and P. Anderson, *Nature*, 2019, **575**, 75–86.

35 R. Ma, S. Tao, X. Sun, Y. Ren, C. Sun, G. Ji, J. Xu, X. Wang, X. Zhang, Q. Wu and G. Zhou, *Nature Communications*, 2024, **15**, 7641.

36 L. Cassayre, B. Guzhov, M. Zielinski and B. Biscans, *Renewable and Sustainable Energy Reviews*, 2022, **170**, 112983.

37 M. Huang, M. Wang, L. Yang, Z. Wang, H. Yu, K. Chen, F. Han, L. Chen, C. Xu, L. Wang, P. Shao and X. Luo, *Nano-Micro Letters*, 2024, **16**, 207.

38 M. Beak, J. Park, S. Park, S. Jeong, J. Kang, W. Choi, W.-S. Yoon and K. Kwon, *Journal of Hazardous Materials*, 2022, **425**, 127907.

39 L. Zhang, H. Gao, Y. Zhu, I. Tran, W. Tang, J. Lin, A. U. Mu, J. Wu, W. Li, D. Nordlund, L. Mu and Z. Chen, *Advanced Energy Materials*, 2025, **15**, 2406084.

40 Z. Meng, X. Ma, J. Hou, Y. Zheng and Y. Wang, *Advanced Energy Materials*, 2025, **15**, 2405383.

41 H. A. Kazem, M. T. Chaichan, A. H. A. Al-Waeli, K. Sopian, W. E. Alnaser, L. Kazmerski and N. W. Alnaser, *Global Energy Interconnection*, 2025, **8**, 363–393.

42 Z. Liu, C. Zhang, M. Ye, H. Li, Z. Fu, H. Zhang, G. Wang and Y. Zhang, *ACS Applied Energy Materials*, 2022, **5**, 14323–14334.

43 B. Zhang, Y. Xu, D. S. Silvester, C. E. Banks, W. Deng, G. Zou, H. Hou and X. Ji, *Journal of Power Sources*, 2024, **589**, 233728.

44 H. Huang, M. Chen, Y. Zhang, X. Wang, Q. Xie, Y. Pu, Y. He, L. Zhu, Y. He and J. Guo, *Langmuir*, 2025, **41**, 1400–1410.

45 P. Ge, B. Wang, Z. Zeng, C. Zhu, H. Lei, J. Li, H. Zhou, W. Sun and Y. Yang, *Advanced Functional Materials*, 2025, **35**, 2500117.

46 H. Lei, P. Ge, Z. Zeng, X. Cui, B. Wang, Y. Yang, X. Ji and W. Sun, *Energy Storage Materials*, 2024, **71**, 103610.

47 Z. Zeng, P. Xu, J. Li, C. Yi, W. Zhao, W. Sun, X. Ji, Y. Yang and P. Ge, *Advanced Functional Materials*, 2023, **34**, 2308671.

48 W. Wu, Y. Lin, Y. Hu, Z. He and Y. Yang, *Journal of Energy Chemistry*, 2025, **102**, 628–650.

49 X. Bai, Y. Sun, X. Li, R. He, Z. Liu, J. Pan and J. Zhang, *Electrochemical Energy Reviews*, 2024, **7**, 33.

50 Z. Yang, Z. Lv, S. Li, J. He and J. Song, *Journal of Environmental Chemical Engineering*, 2025, **13**, 118089.

51 M. Li, R. Mo, A. Ding, K. Zhang, F. Guo and C. Xiao, *Energy Materials*, 2024, **4**, 29.

52 G. Shi, J. Wang, S. Zhang, J. Cheng, X. Shao, Z. Xu, X. Chen and B. Xin, *Separation and Purification Technology*, 2024, **335**, 126144.

53 X. Lai, Q. Chen, X. Tang, Y. Zhou, F. Gao, Y. Guo, R. Bhagat and Y. Zheng, *eTransportation*, 2022, **12**, 100169.

54 A. M. Domingues and R. G. de Souza, *Next Sustainability*, 2024, **3**, 100032.

55 H. Yang, X. Hu, G. Zhang, B. Dou, G. Cui, Q. Yang and X. Yan, *Waste Management*, 2024, **178**, 168–175.

56 E. Crenna, M. Gauch, R. Widmer, P. Wäger and R. Hischier, *Resources, Conservation and Recycling*, 2021, **170**, 105619.

57 S. Du, F. Gao, Z. Nie, Y. Liu, B. Sun and X. Gong, *Journal of Cleaner Production*, 2022, **340**, 130798.

58 J. Sohn, P. Kalbar, B. Goldstein and M. Birkved, *Integrated Environmental Assessment and Management*, 2020, **16**, 314–323.

59 J. Weinzettel, K. Steen-Olsen, E. G. Hertwich, M. Borucke and A. Galli, *Ecological Economics*, 2014, **101**, 115–126.

60 W. Pan and Y. Teng, *Renewable and Sustainable Energy Reviews*, 2021, **141**, 110840.

61 M. Liu, G. Zhu and Y. Tian, *Green Carbon*, 2024, **2**, 425–437.

62 N. Khanna, J. Wadhwa, A. Pitroda, P. Shah, J. Schoop and M. Sarıkaya, *Sustainable Materials and Technologies*, 2022, **32**, e00413.

63 O. Jolliet, M. Margni, R. Charles, S. Humbert, J. Payet, G. Rebitzer and R. Rosenbaum, *The International Journal of Life Cycle Assessment*, 2003, **8**, 324–330.

64 M. G. Rasul and L. K. R. Arutla, *Energy Reports*, 2020, **6**, 503–508.

65 Y. Pamu, V. S. S. Kumar, M. A. Shakir and H. Ubbana, *Materials Today: Proceedings*, 2022, **52**, 1968–1978.

66 C. Mutel, *Journal of Open Source Software*, 2017, **2**, 236.

67 K. Subramanyan, Y. Wu, U. M. Diwekar and M. Q. Wang, *The International Journal of Life Cycle Assessment*, 2007, **13**, 278–285.

68 B. M. Tragnone, M. Serreli, I. Arzoumanidis, C. A. Pelino and L. Petti, *The International Journal of Life Cycle Assessment*, 2023, **28**, 1031–1053.

69 S. Evangelisti, P. Lettieri, D. Borello and R. Clift, *Waste Management*, 2014, **34**, 226–237.

7

Prospects and Outlook

Zihao Zeng, Chao Zhu, Hai Lei, Run Ren, Quan Zhang, Jiale Zhang, Peng Ge

In the future, for achieving greener and higher-quality development of spent lithium-ion batteries (SLIBs) recycling, the relative exploring activities should be continuously carried out. The significant achievements would be expected in resource recycling efficiency, environmental friendliness, and economic value. The detailed contents are listed as follows.

7.1 Pretreatment Technologies

Firstly, for detection technologies, they should be further explored toward efficient, precise, and intelligent battery testing, which is expected to provide comprehensive technical support for the subsequent green and high-value processing of batteries. Herein, the important development directions could be divided into three kinds: intelligent detection, efficient detection, and standardized detection. For intelligent detection, through the establishment of an intelligent battery detection system, it was expected to be coupled with multiple detection methods and data analysis technologies, finally achieving real-time collection, processing, and analysis of detection data. With the support of artificial intelligence (AI) algorithms, the health status and potential risks of the battery could be automatically identified, significantly improving the efficiency and accuracy of detection. For efficient detection, the high-precision and high-sensitivity detection equipment are urgent to explore, meeting the demand of the quick and accurate evaluation of the safety, SOH, and lifespan of the battery. For instance, with the assistance of X-ray imaging, infrared thermal imaging, and electrochemical impedance spectroscopy, the internal state of the battery could be achieved with nondestructive behaviors. Especially the detection time, which could be reduced from several hours to a few minutes, meeting the requirements of large-scale industrial detection. Of course, in order to improve the universality of detection manners, the standardized process should be established, including standardized detection procedures and norms, facilitating the reliability and consistency of detection results. And, the standard process should be composed of practical processes, detailed data, and results evaluations, which would give some significant evaluations for the subsequent utilization (cascade utilization, element extractions, and direct regeneration).

Secondly, for PACK packages, the relative technologies should be explored toward intelligent and refined disassembly. Along with the development of fast identification systems and the QR codes of spent batteries, they could be effectively classified (based on battery categories). Certainly, owing to different designing traits, the practical disassembly should be investigated for AI, multi-structure compatibility, and fully automatic operations. As expected, in the future, all of the mainstream PACK structures can be automatically, efficiently, and precisely disassembled by AI-driven robots. Meanwhile, the automatic recognition and classification accuracy rate would reach up to 100%, accompanied by full-process intelligence (~100%). Moreover, the relative disassembly time is also important, where the evaluation of standard PACK packages should be lower than 1.0 minute. Based on the discussion above, two investigation directions should be noted, mainly containing identifying technologies and the disassembly systems. For the former, the rapid and automated identification technologies should be developed to define the different structures of spent batteries. For the latter, the automatic and intelligent PACK disassembly system should be established, which was mainly composed of modular mechanical arms and a flexible fixture system, contributing to the disassembly of PACKs from multiple brands.

Thirdly, for battery cell, the relative technologies should be investigated toward in-depth discharge and the utilization of residual energy, which is important for further crushing and in-depth disassembling. For the discharging process, the safe discharge is expected, always accompanied by green and highly efficient. Moreover, during discharging, pollution and pollution discharging should be avoided, that is to say, achieving "zero pollution and zero emissions." Thus, the automated deep discharge system should be developed. Meanwhile, multipath energy utilization manners should be integrated to maximize the energy utilization rate. About the discussed technologies above, both discharge depth and energy utilization rate should be up to 100%. To achieve this goal, the key tasks were summarized as follows: the upgrading of the online detection algorithm for residual battery capacity, the development of an energy classification management platform to realize intelligent assessment, and the allocation of different capacity series battery cells. At the same time, further integrating the detection and in-depth discharge platforms, diverse and efficient energy conversion paths would be constructed, with the assistance of thermal/electromechanical energy conversion modules.

Fourthly, for the crushing and sorting of different batteries, the related technologies should evolve toward high safety and efficient directions. Note that a series of advanced technologies should be integrated, such as AI, machine vision, and adaptive robotic arm linkage controlling, enabling the establishment of an intelligent and precise battery safety crushing and an efficient sorting system. Supported by the technologies above, human intervention would be significantly reduced, along with the obtaining of high-purity battery powder. Of course, in the future, both ultrafine powder, copper foil, and aluminum foil should exhibit the recovery rate of over 99.5%; meanwhile, the impurity content in the ultrafine powder is lower than 50 ppm. For achieving this goal, the core tasks should be explored as follows. For instance, the intelligent robot technology should be developed and upgraded, which is conducive to enhancing the precision disassembly and sorting capabilities of retired battery powder. Meanwhile, the AI and adaptive robotic arm linkage control systems should be integrated with detection technologies, further forming a full-process linkage system. The technologies above are anticipated to enable automatic identification of impurities and precise partitioned removal, thereby further promoting the precision level of the disassembly and sorting process of retired battery powder.

7.2 Cascade Utilization

Firstly, the existing challenges should be summarized, providing directions for the related research. Although cascade utilization displays significant economic and environmental benefits, there are still multiple challenges in its practical application. Firstly, it should be acknowledged that diverse SLIBs delivered different capacities, internal resistances, and voltages, resulting in inferior consistency. And, limited by the issues above, they are hardly utilized in cascade utilization. Secondly, potential safety hazards should be emphasized, such as thermal runaway, short circuits, etc., threatening the stability of the system. Finally, the current technological maturity is insufficient, and substantial breakthroughs are still needed in the development of battery reconfiguration and management systems.

Secondly, for the further evolution of cascade utilization, some key tasks should be significantly carried out. In order to solve these challenges above, the cascade utilization technologies should be further developed through the advancement of three major tasks, mainly including sorting technologies, clear failure mechanisms, and effective battery reconfiguration. For developing efficient battery sorting technologies, the precise detection and classification should be further explored to enhance the consistency of battery packs. For a clear failure mechanism, in-depth research should take place, especially on the relationship of failure issues with better practical data, which is beneficial for developing advanced battery management systems (BMS) with real-time monitoring capabilities to prevent safety accidents. Thirdly, optimizing new battery reconfiguration technologies, the system performance and stability should be improved, with the assistance of reasonable arrangements and connection methods.

Thirdly, in the future, an efficient, safe, and sustainable recycling system for retired LIBs is to be built, leading to the formation of a complete industrial chain. Over the next 5–10 years, with the high consistency of the battery group (exceeding 95%), is aimed to be the precise sorting and recombination technologies should be achieved. Furthermore, with the assistance of reliance on real-time monitoring and data analysis, the exploration of an intelligent dedicated BMS should be developed, enabling full-lifecycle management of batteries, thereby optimizing usage efficiency and extending lifespan. Concurrently, to meet the demand of technological innovation and market demand, application fields should be actively expanded. Thus, the large-scale application of retired batteries would be carried out in numerous fields, such as grid energy storage, renewable energy grid connection, and intelligent microgrids. Meanwhile, the automated and intelligent efficient battery system reconfiguration technologies would be developed. This is expected to improve reconfiguration efficiency and system integration capabilities, reduce reconfiguration costs, and enhance economic benefits. Ultimately, the cascade utilization should be carried out with the large-scale and commercial applications, especially in energy storage and backup power supplying, resulting in the formation of a mature and economically feasible closed-loop recycling chain.

As one important member of green and high-value recycling technologies about SLIBs, its developments serve an important role in the efficient utilization of resources and the sustainability of the environment. In the next 5–10 years, about cascade utilization, their evolution would be developed toward higher consistency, higher safety, and large-scale application. Through their effective development, it is expected to achieve efficient recovery and utilization of SLIBs, providing strong support for the large-scale application in their recycling.

7.3 Metallurgical Recycling

Attracted by the global goal of carbon neutrality and the wave of the circular economy, the recycling of SLIBs has advanced from the stage of technical exploration to the critical period of industrialization practice. However, the existing wet and pyrometallurgical processes are still restricted by a series of issues, such as inferior metal recovery efficiency, serious environmental costs, and poor performance of recycled materials. In the future, for the ultimate goal of green and high-value recycling, the limitations of technologies should be solved. Meanwhile, the relative industrial value chain should be further reconstructed. This is to say, the vital elements from recycling industries should serve important roles in the supply of LIBs. Herein, the recycling manners could be divided into hydro- and pyro-manners, where the relative prospects and outlook are listed as follows.

7.3.1 Pyrometallurgical Recovery Technologies

Benefitting from its strong adaptability to raw materials and large-scale processing capabilities, pyrometallurgical recovery is irreplaceable, especially for complex-component battery recycling. However, it should be acknowledged that they still suffer from the high energy consumption and serious lithium loss in the traditional recycling process. For further improving the pyrometallurgical technologies, they would be developed toward low-carbon, intelligent, and high-value, thus the incorporation of multi-technologies would be expected in the future. Especially about the extraction of elements, through precise regulation of physical and chemical processes, the high recovery efficiency may be realized. In terms of energy efficiency improvement, the development of efficient reaction systems, such as oxygen-enriched smelting and plasma smelting, should be devoted to plenty of attention. In the future, the reduction of energy consumption should be up to ~30%. Meanwhile, with the assistance of renewable energy supply, the reduction of carbon emissions should be above 50%.

Meanwhile, for recovering the thermal energy of high-temperature flue gas, the cascade utilization of waste heat should be further promoted. And, about pyrometallurgical recovery, a series of pollutants should be effectively controlled. Firstly, the emission of dioxins should be inhibited, where its emission concentration should be lower than 0.1 ng TEQ/Nm^3. About the control of Pb- and Cd-based pollution from slag, the technologies of directly solidifying heavy metals should be explored, where the leaching toxicity is strictly controlled below the GB 5085.3 standard. Moreover, for refined recycling of metal elements, alloy-targeted separation would capture plenty of attention, and the purity of cobalt-nickel alloys increased up to 99%, with the optimized ratio of reducing agents. About the enrichment of rare metals, the smelting manner should be incorporated with vacuum distillation, achieving a lithium-manganese recovery rate of >90%.

Besides, it should be acknowledged that, during sintering, the recovery rate of lithium elements is relatively low (<60%), and the corrosion of the furnace lining could be always resulted by fluorides. Thus, the pretreatment with sodium salt roasting could induce an increase in lithium recovery rate toward 85%. Meanwhile, with the use of zirconia-alumina composite refractory materials ($ZrO_2 \geq 35\%$), the furnace lining life could be prolonged toward 3 years per furnace run. By 2030, the relative goals should be achieved, such as the comprehensive energy consumption of ≤500 kWh/t (37.5% reduction compared to the current 800 kWh/t), cobalt-nickel recovery rate of ≥98%, lithium recovery rate of ≥80%, dioxin emission of ≤0.05 ng TEQ, automation rate of the production line of ≥90%, and the formation of a mature model with a single line capable of processing 50,000 tons per year. The utilization of developed technologies and a comprehensive revenue of $300/ton should be generated, serving as core roles in the global battery material recycling industry.

Certainly, collaborative innovation should be carried out in the process chain. Combining hydro- and pyro-manners, such as high-temperature crushing incorporated with wet leaching tailings should be explored, where the acid consumption and the amount of solid waste should be reduced about ~40 and ~60%, respectively. In addition, the waste electrolyte may be innovatively used as a reducing agent for the reduction of processing cost ($150/t) and CO_2 (1.2 t/t). For the high-value utilization of waste solid materials, the slag could be used to prepare fire-resistant building materials with a melting point of >1,580 °C, facilitating the increase in value ($50/t) without other waste.

In summary, the lithium battery recycling industry could be promoted by the dual evolution of pyrometallurgical extraction. For metal element recovery, the new-type technologies should be explored for the high extraction efficiency, such as plasma smelting, microwave reduction, and gradient condensation technologies. For pollution protection, a complete clean production system should be established, including oxygen-enriched combustion, carbon capture, and intelligent control. The developments of technologies above are beneficial for the lowering of energy consumption, the promotion of efficiency, and the reduction of heavy pollution. Meanwhile, the new industrial form would be constructed from energy self-sufficiency to resource recycling further to "green emissions." As-known, some innovative technologies would be applied on a large scale in the leading industries (like Tesla, CATL, etc.). By 2030, the global capacity of pyrometallurgical recovery is expected to occupy 40% of the market share, promoting a reduction of more than 60% in the carbon footprint of the entire lithium battery life cycle, bringing about the green future.

7.3.2 Hydrometallurgical Recovery Technologies

As one member of the element extracting technologies, hydrometallurgical recovery has been regarded as the primary recycling technology, mainly ascribed to the limited collection scale of SLIBs. Meanwhile, the traditional manners displayed a long-process treatment, accompanied by serious pollution. In the future, green and intelligent technologies should be explored. Currently, with the assistance of acid leaching and stepwise extraction, hydrometallurgical recovery of elements could be carried out. Although high-value metals such as cobalt and nickel could be successfully extracted, large reagent consumption and difficult wastewater treatment should be emphasized, meanwhile, with a low rate of lithium recovery. In the future, the new reaction systems and intelligent control systems should be integrated in-depth, especially in efficient leaching and separation, targeted purification and impurity separation, and short-process metallurgical and materials.

For the efficient leaching and separation, they served important roles in the extraction of elements from SLIBs. For diverse cathodes, their reaction activity and leaching conditions are clearly different during hydrometallurgical recovery. And, about lithium iron phosphate (LFP), limited by their stability of architecture, the relatively low leaching rate could be found under conventional acid leaching conditions. About NCM materials (metal oxides), the leaching process is easily carried out, but the leaching ratio of Ni/Co/Mn may be tailored, affecting the subsequent separation. Thus, in the future, efficient leaching technologies should be explored for diverse cathodes. For spent LFP, through the utilization of thermal and non-thermal effects of microwaves, microwave-assisted acid leaching technology may be suitable, bringing about the obvious increase in leaching rate, where the leaching rate of lithium can reach up to 95% above. In terms of separation technology, developing new-type extractants and ion-exchange resins should be explored for the efficient separation of metal ions such as nickel, cobalt, and manganese. Based on the previous reports, the phosphorus-based extractant displayed high selectivity for cobalt ions, which could induce the effective separation of Co-ions in complex metal-ion solutions.

For targeted purification and impurity separation, it is well-known that the impurity displayed a serious effect on the recovery of metal elements. And, in the leaching solution, the impurities (such as iron, aluminum, copper, etc.) would be residual, resulting in the lowering of the purity of the product, especially in the recycling of NCM and LFP. About the traditional impurity removal, the chemical precipitation and pH adjustment are always applied. But, owing to the lack of selectivity, the loss of target metals (such as Ni, Co, and Mn) always results. For solving the issues above, the targeted impurity removal technologies should be developed for removing impurities and reducing the loss of valuable metals. For example, with the use of selective complexation precipitants, the relative impurities could be successfully removed under low pH conditions, avoiding the loss of target metals. Additionally, the ion-imprinted polymers (IIPs) with specific recognition capabilities could be developed. Herein, through the matching of pore size and functional group, the relative impurities, such as Fe, Al, and Cu elements, could be successfully captured, with a considerable recovery efficiency. Through the use of new-type technologies, the recovery efficiency could be improved, and the risk of introducing impurities into the target metals can be significantly reduced.

For short-process metallurgical and materials, the incorporation of related technologies in different fields should be emphasized. The traditional hydrometallurgical recovery is composed of metal leaching, solid–liquid separation, impurity removal, metal separation and purification, salt conversion, and material re-synthesis. Especially in multi-metal systems such as Ni/Co/Mn elements, the coordinated development of resource recovery and reuse is seriously restricted by the separation accuracy and conversion efficiency. To achieve process integration and increase value-added, it is urgent to construct an integrated short-process recovery technology, from leaching to regulation, to co-precipitation, and further to precursor synthesis. Benefiting from the effective control of leaching composition and impurities, the co-precipitation of multiple metals could be achieved for the formation of precursors, eliminating other processes, such as separation, extraction, and metal-salt synthesis. For example, by adjusting the adding sequence of pH, complexing, and precipitating agents, Ni/Co/Mn-elements could be prepared for NCM precursors. Moreover, through the shielding strategy of Fe-ions, the growth path of the LFP crystal nucleus could be constructed. As-known, the related technologies were beneficial for the shortening of the recycling process, the reduction of energy consumption, and operational difficulty. Also, through the control of precursors, the particle size distribution, crystallinity, and component uniformity may be effectively tailored, which is beneficial for the obtaining of different precursors. And, with the utilization of continuous reaction equipment and intelligent control systems, the stable control of product quality could be achieved through the adjustment of parameters.

Compared to pyro-recycling manners, the hydro-method displayed low energy consumption, lithium loss, and effective pollution control. In order to solve the issues above, the cooling roasting and intelligent smelting should be integrated, accompanied by secondary extraction of lithium from slag technologies, further achieving efficient collaborative smelting of used lithium batteries and ores. By 2040, the recovery rates of Ni/Co/Cu elements would remain at 99% above, along with the recovery rate of Li-elements (>90%), where the recovery energy consumption would be reduced by 30%. Benefitting from the establishment of short-process metallurgical and materials, the incorporation of precise separation of multiple metals, efficient lithium recovery, and impurity removal should be carried out to form the new-type cathodes, along with the obvious decrease in the treating process. By 2040, the metal recovery rate (Ni, Co, Mn, Li) will be ≥99%, meanwhile the lithium recovery rate will be larger than 95%, and the wastewater reuse rate will reach 100%.

In order to achieve the green and intelligent transformation of SLIBs recycling, their evolution should be triggered. Through innovative integration of efficient leaching and targeted separation, target metals are precisely extracted, whilst impurities are deeply removed. Integrated metallurgy-material technology enables the direct synthesis of a fresh cathode from recycled cathode materials, significantly shortening production cycles and reducing energy consumption. Ultimately, the "zero-waste synergy" ecological closed-loop would be established, achieving near-total metal recovery and complete wastewater reuse. Utilizing cross-industry waste collaboration, the "waste-treats-waste" network would be completed, supporting global recycled resource systems at costs and carbon emissions far below mining levels, while reconstructing the low-carbon circular industry chains.

7.4 Direct Regeneration

Of course, direct regeneration has been regarded as next-generation recycling manners, accompanied by the considerable environmental and economic value, in comparison to element-extracting manners. It should be acknowledged that the traits of raw materials (especially purity) served important roles in their regeneration properties. That is to say, the development of the direct regeneration manners mainly relied on the evolution of their pretreatment process. Moreover, based on the different kinds of materials, their relative prospects and outlook should be divided into spent LFP, NCM, and LCO.

7.4.1 About the Regeneration of Spent LFP

Firstly, to achieve effective restoration of spent LFP electrode sheets without recycling, the significant heterogeneity of physical and chemical properties should be addressed, such as particle size, carbon layer thickness, and diverse doping elements from different sources. And, the short-process regenerated manners should be developed with low-energy consumption. Meanwhile, the solid phase should be emphasized, avoiding the generation of wastewater. Certainly, with the performance of the regenerated materials meeting the demand of commercial requirements, such as capacity, rate, density, etc., for further achieving commercial application. Secondly, considering the spent LFP materials always come from different industries, they displayed diverse physical–chemical properties with failure issues. Thus, to solve their different properties, the technologies formed above should be used in standard manners, accompanied by the development of pre-lithiation technology. As expected, the relative uneven Li-distribution and Li-Fe anti-sites would be solved. Significantly, the properties of as-regenerated samples should meet the demand of commercial materials. Thirdly, in the future, LFP and LMFP may be mixed to pursue considerable electrochemical properties. But, it should be acknowledged that the mixed materials would be hardly repaired. That is to say, if the materials were successfully regenerated, the element content of regenerated samples was different, which would not meet the demand of commercial materials, especially about great consistency. Based on the previous presorts, it should be summarized that the content of Mn-elements served as important roles in their regenerated electrochemical properties. It should be expected that, with the development of the technologies above, the relative cost would be obviously reduced (50%), in comparison with leaching manners. Moreover, the relative carbon emissions and energy consumption could be reduced about 50%, 70%, respectively. Certainly, the great practical technologies always come from the clear and significant theoretical research, thus the relative moving path of vital elements should be illustrated.

7.4.2 About the Regeneration of Spent LCO

The direct regeneration of spent LCO cathodes faces fundamental challenges stemming from extreme feedstock heterogeneity, including particle size variations (0.1–500 μm), inconsistent dopant distribution (e.g., Mg/Al/Ti/Zr/F), diverse coatings (e.g., $MgO/Al_2O_3/ZrO_2/TiO_2/Li_3PO_4$/carbon), and varying degradation levels (20–80% capacity fade). Meanwhile, conventional hydrometallurgical/ pyrometallurgical processes are subject to extreme environmental and economic limitations due to the utilization of heavy acid/alkalis and high temperatures Thus, future technological evolution must prioritize the development of an integrated all-dry regeneration process that eliminates aqueous reagents while enabling precise control over material reconstruction. This necessitates breakthroughs in three interconnected domains: overcoming feedstock complexity via advanced dry preprocessing, mastering precision dry repair/re-lithiation, and integrating system sustainability.

First, future advancements need to overcome feedstock complexity via advanced dry preprocessing. Specifically, prioritized ultraefficient dry processing chains should be designed to handle the immense variability inherent in spent LCO, and the key developments are as follows. Next-generation dry grinding: Developing highly efficient and energy-optimized dry milling techniques (e.g., advanced jet milling and impact milling with precise control) capable of achieving consistent sub-micron or even nanoscale particle sizes without inducing detrimental phase changes or excessive heat. Besides, intelligent dry classification and sorting: Integrating advanced dry classification methods (e.g., multistage classifiers, electrostatic separation, and triboelectric separation) combined with real-time sensing (in situ Raman and XRF) to achieve incomparable separation precision. It should be noted that this targets not just particle size, but also the removal of contaminants (Al, Cu foil fragments, binders, and carbon), isolation of specific coating types, and potentially even grouping particles by degradation level within the LCO stream.

Second, precision dry repair/re-lithiation should be mastered in the future. The core of future direct regeneration lies in defect-targeted, highly controlled solid-state reactions, which necessitate moving far beyond simple high-temperature roasting. This encompasses tailored reaction engineering, which involves designing reactor systems and thermal profiles that provide spatial and timing control over temperature, atmosphere (O_2 partial pressure, gas flow), and potential mild mechanical activation. Processes must be adaptable to accommodate the specific degradation state identified during classification. For instance, lower temperature/shorter time for mildly degraded material and optimized conditions for severely Li-deficient or structurally damaged particles. In addition, advanced solid-state re-lithiation should be adopted to perfect the uniform dispersion and reaction kinetics of solid lithium sources (e.g., Li_2CO_3 and LiOH) within the milled LCO mixture under strictly dry conditions. This includes exploring mechanisms like eutectic melting or enhanced solid-state diffusion promoted by carefully controlled microenvironments. Furthermore, defect-specific reconstruction entails developing deep understanding and control over the solid-state processes that repair cation mixing (Li/Co disorder) and oxygen vacancies, without leading to cobalt reduction or oxygen release. This may involve novel catalytic surface treatments or precisely timing atmosphere swings.

Third, system sustainability requires integrating environmental and economic viability from the outset. This involves focusing on three key aspects: closed-loop fluorine management, which implements robust dry capture and conversion systems for fluorine (from PVDF/$LiPF_6$ decomposition) into stable, nontoxic compounds (e.g., CaF_2 and AlF_3) suitable for safe disposal or potential reuse, eliminating gaseous HF risks. Radical energy efficiency, minimizing energy consumption through process intensification, heat integration, and potentially utilizing renewable energy sources. And, low-temperature pathways (<600 °C) will be a major goal. Adaptive process control and AI, which utilize advanced process analytics (in-line monitoring of temperature, gas composition, and material structure) coupled with machine learning for real-time feedback control, ensuring consistent output quality and fully automated operation. Material design for regeneration, fostering collaboration between cell manufacturers and recyclers to design future LCO cathodes with regeneration in mind (e.g., markers for degradation state).

In conclusion, the future of spent LCO regeneration is unequivocally dry, intelligent, and precise. Supported by mastering ultrafine dry homogenization, intelligent classification, and defect-specific solid-state reconstruction within highly controlled reactor environments, this technology can achieve universal applicability across the full spectrum of spent LCO grades, exceptional environmental performance, and economic viability. This transformation is essential for establishing truly circular and sustainable LIB ecosystems.

7.4.3 About the Regeneration of Spent NCM

Differing from the traits of LFP and LCO, NCM samples displayed a series of kinds, containing NCM523, NCM622, NCM811, and so on. That is to say, the collected spent materials were composed of diverse materials, which brought about the serious regeneration process. Moreover, in order to pursue great electrochemical properties, some coating layers (like Al_2O_3, ZrO_2, etc.) and heteroatoms (like W, B, etc.) were always introduced to induce the tailoring of their physical–chemical properties. Thus, for effective regeneration of spent NCM, their significant regeneration technologies should be further explored. Thus, in the future, the single samples should be divided into single crystal, multi-crystal, mixed spent materials, and so on. Given this, according to the difference of spent materials, the relative exploring content was shown as follows.

Firstly, for electrode about spent single crystal without liquid injection, the related technologies and mechanisms should be explored for solving surface/near-surface damage from the removal process of slight carbon content and binder. Especially for PVDF binders, the existing F-elements served important roles in their electrochemical properties, perhaps leading to the fading of their physical–chemical traits. And, considering their different element compositions, their changing process and distribution state should be clearly explored. Meanwhile, their effect on the energy-storage abilities would be further illustrated. And, based on the rapid developments of AI, the relative threshold value of impurity elements and the elemental formulas should be solved.

Secondly, for the multi-crystal NCM materials, the issues above should also be solved. Meanwhile, it should be emphasized that the ball milling would render the dissociation of samples from big size to small size. Thus, during regeneration-sintering process, the growth of small particles should be deemed as the next-generation exploring direction, containing the growth process, sintering conditions, and product traits. Obviously, they would serve important roles in their electrochemical performances. Meanwhile, owing to the introduction of ball milling, the strong mechanical forces perhaps result in further damage of particles, bringing about further injury to spent materials, especially microcracks, surface physical damage, and so on. As-known, their recovery may be the new-type challenges. In addition, owing to the different kinds of NCM, their uniform mixing state would determine the final state of regeneration materials. That is to say, limited by the sluggish diffusion kinetics, the excellent single crystal is hardly prepared, whereas NCM95, NCM811, and NCM622 may be produced at the same time. For further commercial applications, the product with different traits hardly meets the demand of industry, especially product stability about different batches. Of course, in order to solve the issues above, the molten-salt growth method should be emphasized, which could offer the ion-diffusion matrix in high-temperature ionic liquids, such as LiOH-$LiNO_3$ systems, LiOH systems, and so on. But, considering the relative cost of molten-salt systems, the low-cost systems should be explored in-depth for commercial productions. Moreover, the relative cosolvents may also be suitable for the fusion of different grains. Considering their relatively low usage of content, the relative manners would be promising potential. As-known, about the preparation of commercial single-crystal particles, solid-state manners, and liquid-state manners are always used from precursors to commercial materials, mainly ascribed to the rich ion-insertion channels. However, for the preparation of single crystals from spent NCM, the solid-sintering manners may be suitable due to the relatively integrated crystal structure. Therefore, captured to the low cost of solid-sintering manners, it has been deemed as the next-generation regeneration manners.

Thirdly, for spent NCM from failed batteries after cycling, the difficulty of regeneration manners was clearly increasing. Based on their failure mechanism, it could be divided into elements-dissolution, lithium loss, too-excess CEI formation, and so on. Perhaps, the single preparation manners hardly induces the successful regeneration of spent materials, especially the significant recovery of all defects. With the developments of single-crystal growth manners, the relative issues above are expected to be solved. But, some impurities should not be neglected, such as serious F-element erosion, and the residual F-residuals (CEI films, electrolyte, etc.). Moreover, owing to the serious electrolyte corrosion, the Al impurities would be largely introduced, which perhaps resulted in the inferior electrochemical properties. Thus, in the future, the relative removal manners of impurities should be emphasized for the significant regeneration of spent NCM.

In conclusion, in the future, the relative exploration of spent NCM regeneration should focus on the removal of impurities and the fusion of serious fading materials. Of course, the AI technologies would be beneficial for the rapid progress about the regeneration manners, contributing to the high-entropy strategies and single-crystal fusion manners. Of course, during regeneration, the relative diffusion and distribution of hetero-atoms should be further explored. As anticipated, with the solving of issues above, the successful regeneration should be expected for the green closed-loop recycling of NCM.

7.5 About the Pollution Controlling and All-life Cycling Assessments

As-known, during the recycling of spent batteries, pollution was always existing in all processes, such as crushing, element extraction, regeneration, and so on, which would bring about gas pollution, spent salt, etc. And, based on the previous work, without significant recycling, the utilization of LIBs has been regarded as inferior to green fields. Thus, the exploring of all life cycle assessments is important for the use of LIBs. More importantly, for LIBs, the building of low-carbon and closed-loop recycling system is key for achieving eco-friendly development throughout the entire battery life cycle. Of course, in order to meet the demand of strategic goals, four points should be achieved, containing the minimizing of environmental risks, the maximizing of resource recycling, the deep reduction of carbon footprints, and the balancing of economic with ecological benefits. And, for achieving the targets above, a series of actions should be completed as follows. About pollution prevention, the fluorine-containing pollutants and multistate wastes should be solved at the same time, especially the fully enclosed exhaust purification, high-value use of solid waste, and closed-loop regeneration of wastewater. Meanwhile, for low-carbon processes, some effort should be devoted to replacing traditional leaching solvents, optimizing intelligent sorting systems, and upgrading of safe crushing procedures, further bringing about the reduction of energy and emissions. In addition, the dynamic databases and blockchain traceability technology should be integrated for life-cycle management, further preparing multi-objective decision-making models to guide process optimization. And, during the life-cycle assessments, the assistance of multi-agent cooperations was required, containing the dedicated material circulation channels, financial incentives, and transparent community environmental data, achieving the green recycling of SLIBs. Finally, about the highly effective recycling of SLIBs, the simple waste treatment should be changed into value creation. And, the sustainable and eco-friendly recycling systems should be explored with considerable economic value toward battery recycling. Thus, the relative mechanism and technologies should be carried out as follows.

Firstly, the clean recycling technologies should be promoted in all the processes, especially about the generic technologies, such as the gas purification, the waste heat utilization, the "zero" wastewater discharging, the green leaching with separation techniques, and the direct repairing technologies. Certainly, with the formation of technology chains, they would be widely applied in industry. More significantly, in the future, the recycling production lines should be optimized, supported by the introduction of excellent technologies with highly effective environmental protection. And, the source pollution reduction and process controlling should be combined, contributing to the further reduction of pollution, while the recycling rate of vital elements is expected to achieve 100%.

Secondly, the emission production of waste liquid and slag should be obviously reduced. Regarding the full-process controlling, it could be divided into three parts: the reduction of pollution at the sources, closed-loop recycling in the process, and the high-effective treatment at the end. For example, through the introduction of new-type leaching with low residuals, bio-based extractants, and so on, the emission reduction of waste liquids and slags could reach up to 80 and 90%, respectively. Meanwhile, the utilization manners of sulfate by-product and fluorine-containing wastewater should be further explored, rendering the successful transformation toward battery-grade or chemical-grade products.

Thirdly, about the all-life cycling assessments, the full-lifecycle monitoring system and closed-loop management mechanism should be established, further tracing the elements and energy diffusion and transformation. Supported by digital environmental monitoring and standardized LCA tools, the environmental performance of different production batches is evaluated at first time, which is beneficial for establishing a comprehensive risk management system from recovering to treating. Meanwhile, the incorporation of recycling and manufacturing industries should be explored at the same time, which is anticipated to achieve the transformation from recycled materials to produce new batteries. Note that, the batch stability should be less than 5%. More importantly, about the designing of batteries, recycling-friendly concepts should be incorporated, especially the simplify dismantling through standardized components. In the end, these measures would ultimately reduce the carbon footprint per kilogram of lithium battery materials by over 70%, while significantly improving resource circulation efficiency.

Based on the above analysis, for achieving green and sustainable recycling of SLIBs, the integrated implementation of clean technologies is required, accompanied by low-carbon processes and full-lifecycle management. Assisted by the establishing of closed-loop systems, the optimizing of resource efficiency, and the incorporation of digital monitoring, the transformation from simple waste treatment to true value creation is expected. This holistic approach, supported by multi-stakeholder collaboration, will ultimately minimize environmental impact while with maximizing economic benefits, paving the way for a genuinely circular battery economy.

8
Conclusion

Chao Zhu, Hai Lei, Run Ren, Quan Zhang, Jiale Zhang, Zihao Zeng, Peng Ge

In summary, based on the discussions above, it could be noted that the recycling of spent displays has considerable economic and environmental value. Certainly, with the significant developments of related technologies, the closed-loop recycling would be achieved, with the reduction of pollution and the utilization of energy and elements, further bringing about the rapid and green achievements of this industry. Thus, the successful recycling of spent batteries is very important in a green and sustainable society. Herein, in detail, six key chapters are prepared in the book, containing pre-treatments, cascade utilization, hydrometallurgy/pyrometallurgy elements extracting, direct regeneration, pollution controlling, and all-life cycling assessments, and their prospects and outlook. About all of them, the recent developments were shown and discussed in depth, meanwhile offering significant future development directions in different fields. Moreover, the relative comparison of different technologies is exhibited, illustrating their advantages and disadvantages.

About the pre-treatments for the future of spent lithium-ion battery recycling, the development of intelligent, integrated, and green processes should be noted, achieving high resource efficiency, environmental friendliness, and economic value. About detection, machine learning would render the evolution of smart and universal systems, which would bring about the rapid and nondestructive battery health assessment. Assisted by the achievements above, the analysis time would be reduced to seconds, where the standardized protocols are established for second-life applications. And, for precise disassembly, AI-driven robotics would enable the evolution of fully automatic recognition and disassembly of all mainstream PACK structures, for targeting 100% accuracy rate, meanwhile with optimized disassembly time (<1.0 minute). About the discharge process, green and safe manners should be exploring, also with the investigation of an automated system, for achieving "zero pollution and zero emission." And, the deep discharge depth of 100% was expected, accompanied by a residual energy utilization rate (>99%).

Finally, in crushing and sorting, the integration of AI and machine vision should be introduced, which is conducive to the evolution of intelligent, safe operations for obtaining high-purity materials. Meanwhile, the recovery rates are expected to reach up to 99.5% above for pole powder, where the content of Cu and Al should be controlled below 50 °ppm. Based on the progress of the technologies above, a sustainable and economically viable battery recycling ecosystem could be established for the highly effective pretreatment.

As-known, cascade utilization is critical for extracting residual value from retired lithium-ion batteries, yet it still suffers from significant challenges. Owing to different failure states, the assembled battery packs would be limited by obvious safety risks, along with the reducing of system efficiency, where the overall capacity is constrained by the weakest unit. Moreover, unstable electrochemical characteristics bring about the increasing risk of thermal runaway. To address these issues, an optimized echelon utilization framework should be established. And, the standardized sorting criteria should be developing for the specific application scenarios. Meanwhile, the intelligent repackaging systems should be implemented, through the advanced clustering algorithms, to improve the consistency of module. Also, the specialized battery management systems (BMS) were required for real-time safety monitoring and heterogeneous compatibility. Furthermore, the relative application fields should be diversified into grid storage, renewable energy integration, and microgrids. Supported by these measures above, the residual capacity could be effectively optimized within a safety governance framework, maximizing lifecycle value, and making echelon utilization an indispensable component of the closed-loop battery recycling value chain.

About the element-extracting process (pyrometallurgy and hydrometallurgy), they are always regarded as core recycling methods. A series of technological advancements should be expected for overcoming their respective limitations. About pyrometallurgy, the high energy consumption and substantial lithium loss were regarded as vital challenges. Thus, the efficient, low-temperature smelting systems should be expected, such as the introduction of molten salts, the optimizing of slag chemistry with additives, and the employing of electromagnetic stirring. Certainly, spent batteries and raw ores could be handled simultaneously, contributing to the reducing of energy consumption, and the highly effective extraction of lithium elements. About hydrometallurgy, the less energy-intensive process, it struggles with low lithium recovery rates and complex impurity removal, which still exist, hindering the production of high-purity products. To address the issues above, the dynamic pH control and highly selective adsorbents (e.g., functionalized MOFs) should be developed for preferential lithium recovery. For impurities, the specific complexing agents and ion-imprinted polymers are also crucial. More significantly, the integrating intelligent crystallization could be used to directly synthesize high-purity battery-grade precursors, achieve the closed-loop transformation from recycled materials directly to high-value products, which enhances the overall economic viability of the recycling chain.

Direct regeneration of cathode and anode materials is pivotal for sustainable battery recycling, especially in the distinct challenges of diverse electrode materials. For LFP, systematic breakthroughs are needed to establish a low-cost, efficient industrial chain, especially the enhanced stability of product batches and low energy consumption. Noted that, considering their diverse physical–chemical properties, the relative homogenizing manners should be expected. Meanwhile, for LCO regeneration, the complex composition of elements is the vital limitation; meanwhile, the sluggish ion-diffusion should be expected, along with the obvious reducing of energy consumption. And, the innovative approaches, such as microwave-assisted and molten salt methods, should be further explored, but with the reducing of cost, for the further industrialization of the regeneration of spent materials. Differing from the cathode materials above, spent NCM displayed a complex failure mechanism and physical–chemical compositions. Thus, the relative fusion of different raw materials should be explored. Certainly, the theoretical calculations with in situ characterization should be developed for the selection of a suitable method (e.g., solid-state, hydrothermal, or molten salt). Crucially, developing self-saturating lithium-rich systems is a potential solution for precise re-lithiation. For regenerating spent graphite, their surface and internal defects should be carried out, along with low-energy consumption. And, the low-temperature purification should be combined with high-temperature structural repairing. Also, the intelligent control and continuous production lines are still important. Ultimately, regarding the regeneration manners, the innovation should concentrate on the precise controlling and green processes; meanwhile, system integration is essential to advance direct regeneration toward commercial viability.

About pollution and all-life evaluation, a clean and efficient system should be established. And, the significant strengthening of pollution prevention is required to minimize environmental impacts, further achieving sustainable development at all of the entire life cycle. Moreover, the excellent recycling efficiency is expected, along with zero harmful emissions and zero environmental incidents during the recycling process. For all life cycle assessments, the relative process should be further optimized, ensuring a substantially reduced carbon footprint per kilogram of recycled material in comparison to current benchmarks. The recycling ecosystem should be built, where economic feasibility is combined with environmental friendliness. Especially the utilization of valuable materials like lithium, cobalt, and nickel, the primary resource consumption and pollutant emissions should be minimized. Ultimately, the principles of a circular economy should be fulfilling, such as resource circulation, environmental sustainability, and shared benefit for all process.

Finally, in the future, recycling manners should be developed toward intelligent, green, and integrated technologies. Given this, the book is expected to offer a comprehensive discussion about the recycling of spent batteries during the regeneration process, meanwhile providing significant guidance.

Index

Note: Page numbers in *italics* and **bold** refers to figures and tables respectively.

d